Rebirth of the Sacred

REBIRTH OF THE SACRED

*Science, Religion, and the
New Environmental Ethos*

Robert L. Nadeau

OXFORD
UNIVERSITY PRESS

OXFORD
UNIVERSITY PRESS

Oxford University Press is a department of the University of Oxford. It furthers the
University's objective of excellence in research, scholarship, and education by publishing
worldwide.

Oxford   New York
Auckland   Cape Town   Dar es Salaam   Hong Kong   Karachi
Kuala Lumpur   Madrid   Melbourne   Mexico City   Nairobi
New Delhi   Shanghai   Taipei   Toronto

With offices in
Argentina   Austria   Brazil   Chile   Czech Republic   France   Greece
Guatemala   Hungary   Italy   Japan   Poland   Portugal   Singapore
South Korea   Switzerland   Thailand   Turkey   Ukraine   Vietnam

Oxford is a registered trade mark of Oxford University Press in the UK and certain other
countries.

Published in the United States of America by
Oxford University Press
198 Madison Avenue, New York, NY 10016

© Oxford University Press 2013

All rights reserved. No part of this publication may be reproduced, stored in a retrieval
system, or transmitted, in any form or by any means, without the prior permission in
writing of Oxford University Press, or as expressly permitted by law, by license, or under
terms agreed with the appropriate reproduction rights organization. Inquiries concerning
reproduction outside the scope of the above should be sent to the Rights Department,
Oxford University Press, at the address above.

You must not circulate this work in any other form
and you must impose this same condition on any acquirer.

Library of Congress Cataloging-in-Publication Data
   Nadeau, Robert, 1944–
   Rebirth of the sacred : science, religion and the new environmental ethos /
Robert L. Nadeau.
       p. cm.
   Includes bibliographical references and index.
   ISBN 978–0–19–994236–7 (hardcover : alk. paper)
   1. Environmental ethics.   2. Religion and science.   3. Environmentalism—Religious
aspects.   4. Human ecology—Religious aspects.   I. Title.
   GE42.N34 2013
   179'.1—dc23
   2012014278

ISBN: 978-0-19-994236-7

9 8 7 6 5 4 3 2 1
Printed in the United States of America
on acid-free paper

This book is dedicated to my good friend and
colleague Professor Peter B. Brown.

TABLE OF CONTENTS

Introduction   *1*

1. Rebirth of the Sacred: Science, Religion, and the New Environmental Ethos   10

    A not so modest proposal   13

    The science of global warming   15

    The bad news about global warming   19

    The new terms of human survival   20

    The politics of global warming   22

2. The New Story of Humanity: Brains, Minds, and the Environmental Crisis   25

    The amazing gift of language   29

    A new view of human nature   31

    A brief early history of the human family   34

    Human transformation of the biosphere   36

    Human population growth and the energy regime of oil   38

    Lessons from the new story   42

3. The New Story in Physics: Mind, Matter, and the Nonlocal Universe   45

    The strange new world of quantum physics   49

    A new fact of nature   56

    Lessons from the new story   59

4. The New Story in Biology: Parts and Wholes in the Web of Life   62

    Parts and wholes in the system of life   66

    Competition versus cooperation   69

    Parts and wholes in public policy   72

    Lessons from the new story   73

5. The Old Story: Sovereign Nation-States and Myths of National Identity   76

    Origins of the construct of the sovereign nation-state   77

    The emergence of churches of state   80

    Narratives about nationalism and national identity   82

    The present system of international government   84

    The legal principle of state sovereignty   87

    In false gods we trust   89

6. The Old Story: Metaphysics, Newtonian Physics, and Classical Economics   92

    The system of natural liberty   95

    The cosmic machine and hidden chains   97

    The businessman   99

    Tightening the chains: Thomas Malthus and David Ricardo   100

    The not-so-worldly philosophers   104

7. The Old Story: Metaphysics, Mid–Nineteenth Century Physics, and Neoclassical Economics   107

    The creators of neoclassical economics   110

    A green thumb on the invisible hand   115

8. The Old Story: Economic Globalization, the Market Consensus, and the New State Religion   121

    The God with the invisible hand and the market consensus   125

    In the market we trust   127

    The religion of the market   129

    Toward a new theory of economics   130

9. The Dream of This Place with Us: Science, Religion, and the Environmental Crisis   137

    The new story of the universe   141

    The truths of science and religion   143

    The new environmental ethos   146

    False gods and the politics of global warming   149

Notes   157

Index   171

Rebirth of the Sacred

Introduction

In the house that seems embarrassingly large and mostly empty now that our children are grown and have children of their own, there is a screened-in porch adjacent to the family room on the second floor. When my wife and I sit on this porch during the summer, after the leaves on the old growth trees in our back yard open and many varieties of birds magically appear, it seems, for a moment at least, that all is right with the world. Laurel oaks and Virginia pines tower over a landscape covered in a canopy of intense green leaves, and members of two families of gray squirrels perform acrobatic feats in the maze of limbs and branches. American goldfinch with bright yellow bodies and black wings, northern cardinals with red bodies and orange bills, and pileated woodpeckers with large red crests and moustaches are frequent visitors. After evening comes and the sky begins to grow dark, the magical mystery tour continues as deer and the occasional lone fox or coyote pass through the thick foliage in the woods behind the house.

But when we sat on this porch during the summer of 2010, it was not possible to feel even for a moment that all was right with the world. The outside temperature was oppressively hot for most of the day, and there were record-breaking heat waves across the United States, Europe, and Asia. The temperature in South Central Pakistan in May was the highest ever recorded in Asia, 128 degrees Fahrenheit, and the average temperature in Moscow in July was an unprecedented 100 degrees Fahrenheit. In August, scientists at the World Meteorological Organization (WMO) said that research based on enhanced climate models indicated that global warming was responsible for these much higher than average temperatures.

The scientists at WMO also concluded that higher temperatures in the Atlantic Ocean and atmospheric anomalies caused by a La Nina created conditions that resulted in the severe draught in Russia, the flooding and mudslides in western China, and the floods that inundated about one-fifth of the landmass in Pakistan. The draught in Russia resulted in wildfires that killed fifty-three people, left thirty-five hundred homeless, and destroyed twenty percent of the wheat crop. The flooding and mudslides in China were responsible for the deaths of about twelve hundred people in addition to the destruction of thousands of homes. The floods in Pakistan caused the deaths of at least three thousand people, displaced about twenty million, and left more than six-and-a-half million without access to adequate shelter, food, potable water, and medicine.[1]

There was ample coverage of the suffering of the victims of these natural disasters on the mainstream American news media. But what the commentators on these programs failed to mention is that these so-called natural disasters were linked to man-made global warming. The list of other inconvenient scientific truths that the managers of these news programs did not find newsworthy in the summer of 2010 is impressively long. Nothing was said on these programs about the fact that half the people on this planet live in countries where water tables are falling and wells are going dry. There was not a single mention of the equally disturbing fact that water from the melting of the Greenland Ice Sheet will soon inundate the rice-growing river deltas in Asia and many coastal cities in this region with large populations. The audience for these news programs was not told that disruptions in the seasonal flow of water from mountain glaciers in the Himalayas and the Tibetan Plateau is resulting in crop failures and severe shortages of drinking water in India and China. And those who listened to these programs on radio or watched them on television were not informed that forests are shrinking by thirteen million acres per year and that four-fifths of the oceanic fisheries are fished beyond capacity or overfished to the point where they could soon collapse.[2]

The scientists who published the research containing the bad news about the state of the environment took pains to remind their readers that average earth temperature will continue to increase due to residual warming caused by carbon dioxide already in the atmosphere. And they also stressed that this residual warming would result in dramatic increases in the frequency of crop-withering heat waves, intense draughts, severe flooding, and more powerful and destructive storms. But nothing was said about this dire prediction on the mainstream American news media, and one reason why this was the case is apparent in an email sent by a Fox news executive to editors and reporters on this network. In this email, the executive told the journalists to "refrain from asserting that the planet has warmed (or cooled) in any given period without IMMEDIATELY pointing out that such theories are based upon data that critics have called into question."[3]

Meanwhile, oil and energy companies were funding massive disinformation campaigns on radio and television designed to convince Americans that their economic well-being was utterly dependent on the consumption of increasing amounts of "clean and plentiful" fossil fuels. One pervasive message in this campaign was that imposing an "energy tax" on these fuels would increase unemployment and retard the growth and expansion of the American economy. The other was that drilling for oil off the coasts of the United States, extracting oil from tar sands, and releasing natural gas from shale rock formations with the use of the environmentally destructive process of fracking would reduce dependence on foreign sources of energy and enhance national security.

The slick commercials in this disinformation campaign appeared with great frequency on talk shows on radio and television where political pundits and politicians were preoccupied with the impacts of the economic recession that resulted from the meltdown of the financial markets in September 2008. The expectation on these programs was that those on the one side of the partisan divide between the political parties would blame those on the other for the loss of jobs, houses, and retirement funds and for the failure to stimulate the growth and expansion of the American economy. But commentators at both ends of the political spectrum were in agreement that the light at the end of the dark tunnel of the economic recession was the prediction by economists at the World Bank and the major investment banks that the global market system would expand by three percent a year until the year 2035.

What the economists who made these predictions did not know or chose to ignore is that if the global market system continues to expand at this rate with the use of existing energy resources, global warming will trigger irreversible large-scale changes in the climate system.[4] And if, as now seems likely, these changes occur, the lives of billions of people will be threatened, competition for scarce resource will result in cross-border conflicts and wars of aggression, and the global market system will collapse.

Recent polls suggest that the disinformation campaigns about global warming have convinced many people in the United States and other highly industrialized countries that this problem does not really exist. Anyone who believes that this is the case should spend some time reading reports done by security analysts at the Pentagon on the threats posed by global warming to the national security of the United States. In the introduction to one of these reports, the authors warn that the assumption in the minds of "most people" that climate change will be gradual "may be a dangerous act of self-deception" because "recent evidence suggests that the possibility that a more dire climate scenario may actually be unfolding."[5] They then state that the abrupt climate change scenario that serves as the basis for assessing threats to national security in their report should be taken very seriously because "some recent scientific findings suggest that we could be on the cusp of such an event."[6]

In this scenario, global warming causes additional melting of the Arctic glaciers and increased rainfall, massive amounts of fresh water lower the salinity of the waters in the North Atlantic, and this results in the collapse of a massive current that flows north from the tropics known as the global thermohaline conveyor. The circulation pattern of the Gulf Stream, the northern arm of this conveyor that now carries warm water into northern latitudes, shifts dramatically over the next ten years. The collapse of this massive current causes other large-scale changes in the global environment, and these changes create conditions that vastly increase tensions and conflicts between nation-states.

At the end of this ten-year period, the average temperature in Asia, North America, and Europe decreases by five to six degrees Fahrenheit, and the average temperature in Australia, South America, and southern Africa increases by about five degrees Fahrenheit. Drought exists in most agricultural regions, and there are severe shortages of potable water in most population centers, including those in Europe and eastern North America. In Northern Europe, annual rainfall declines by thirty percent, and the climate resembles that in present-day Siberia. Conditions are so extreme in Scandinavian countries that large numbers of people in this region migrate to warmer and more habitable climes in southern Europe. In Western Europe, winter storms intensify, and the destructive impacts of these storms are amplified by strong westerly winds in the North Pacific. In Africa, climate change causes widespread famine and disease, large numbers of starving immigrants from this region attempt to enter countries in southern Europe, and these countries secure their borders in an effort to stem the tide of unwanted immigrants.

Unpredictable monsoons in China cause devastating floods in areas where vegetation has virtually disappeared due to droughts, and longer, colder winters and hotter summers create severe energy and water shortages. Similar problems exist in parts of Asia and East Africa, and tensions between countries in this region occasion a series of cross-border military conflicts. Violent storms are increasingly common virtually everywhere on the planet, the lives of millions of people are threatened by floods, and most of Bangladesh is uninhabitable due to rising sea levels and contaminated water supplies. Rising ocean waters from one of these storms break through levees in the Netherlands, making coastal cities like The Hague unlivable.

In the United States, storms breach the delta island levees in the Sacramento River, salt water can no longer be kept out of the aqueduct system that carries potable water to large populations in Southern California during the dry season, and this massively disrupts the supply of fresh water to this region. Colder, windier, and dryer weather in northeastern states make growing seasons shorter, and longer dryer conditions in southern and southwestern states dramatically reduce agricultural production in these regions as well. High winds and reduced rainfall in virtually all agricultural areas in the United States result in soil loss and reduced moisture in the soil, and this contributes to additional declines in food production.

Cooler temperatures in the northern hemisphere drive up consumption of oil, and many industrialized countries respond to the large increases in the price of oil by greatly increasing their reliance on nuclear energy. As conflicts between nation-states over access to scarce environmental resources escalate, Japan, South Korea, and Germany use the spent fuel from their nuclear plants to develop nuclear weapons. Japan, menaced by flooded coastal cities and contaminated water supplies and lacking sufficient oil and gas reserves to power its massive desalination plants and energy intensive agricultural system, develops plans to gain access to the nearby Russian oil and gas reserves with the use of military force. If that occurs, the authors predict that the resulting conflict could easily escalate to the point where one of these countries could elect to use nuclear weapons. In another scenario, a series of cross-border conflicts between Pakistan, India, and China escalate to the point where, the authors conclude, an exchange of nuclear weapons would be highly probable.

In this geopolitical climate, say the authors, "Nations with the resources to do so may build virtual fortresses around their countries," and the primary concern of these countries should be to secure "resources for survival" as opposed to defending "religion, ideology, or national honor."[7] In the conclusion of the report, the authors describe some ways in which the United States could build a virtual fortress around itself and some "unlikely alliances" that may be required to secure the resources necessary to sustain the national economy: "Borders could be strengthened around the country to hold back unwanted starving immigrants from Caribbean islands (an especially severe problem), Mexico, and South America. Energy supply will be stored through expensive (economically, politically, and morally) alternatives such as nuclear, hydrogen, and Middle Eastern contracts. Pesky skirmishes over fishing rights, agricultural support, and disaster relief will be commonplace. Tensions between U.S. and Mexico rise as the U.S. reneges on the 1944 treaty that guarantees water flow from the Colorado River. Relief workers will be commissioned to respond to flooding along the southern part of the east coast and much drier conditions inland. Yet even in this continuous state of emergency the U.S. will be positioned well compared to others."[8]

From the perspective of environmental science, the claim that the United States will emerge from this chaos "positioned well compared" to other countries makes no sense at all. The collapse of the global thermohaline system would have disastrous impacts on people living in every region and territory on the planet. And the collapse of this system would trigger a cascade of large-scale "irreversible" changes in environmental subsystems that would have disastrous impacts on all human populations for a very long time. In this scenario, there is no prospect that the government of any nation-state, including that of the last remaining superpower, could protect its citizens, much less sustain a growth economy.

Fortunately, scientific research published after this Pentagon report appeared revealed that the collapse of the global thermohaline system is not as imminent as the earlier research indicated. But this research has also demonstrated that we are perilously close to the point where global warming could trigger irreversible large-scale changes in the climate system that would have impacts on human populations very similar to those described in this report. For the purposes of this discussion, what is most important about the Pentagon report is that it illustrates a large problem that constitutes the greatest barrier to resolving the environmental crisis. The problem is that the political and economic narratives that now serve as the basis for coordinating global human activities are predicated on unscientific assumptions that effectively preclude the prospect of implementing scientifically viable solutions for problems in the global environment.

The political narrative is premised on the construct of the sovereign nation-state, and the only source of political power in the present system of international government, the United Nations, is the sovereign nation-state. The economic narrative is premised on the construct of the invisible hand, or on the belief that the "lawful" dynamics of market systems legislate over decisions made by economic actors and maintain these systems in relative equilibrium. The chapters on the origins and history of these constructs will not only reveal that these narratives are predicated on metaphysically based and scientifically invalid assumptions about the sources of human identity and the relationship between human and environmental systems. They will also demonstrate that there is no basis in these narratives and in the institutional framework and processes associated with the narratives for implementing the scientifically viable public policies and economic programs required to resolve the environmental crisis.

What should become quite clear during the course of this discussion is that if we are to resolve the environmental crisis, two quite remarkable developments must occur in very little time. The first is that the present system of international government must be replaced by a system that is not predicated on the construct of a sovereign nation-state capable of implementing scientifically viable solutions for environmental problems. And the second is that the theory that now serves as the basis for coordinating global economic activities, neoclassical economics, must be replaced by an environmentally responsible theory that can serve as the basis for implementing scientifically viable economic solutions for these problems.

The admittedly very ambitious aim in this book is to make a convincing case that these massive changes in our political and economic institutions could occur if sufficient numbers of environmentally concerned people in the five great religious traditions of the world enter the new dialogue between the truths of science and religion. There are two reasons why I am convinced that this claim has substantive validity. First, this dialogue can serve as the basis for articulating and disseminating an environmental ethos with a profound spiritual dimension that can be embraced by people in all of the great religious traditions of the world. And second, the widespread acceptance of this ethos could result in the fairly rapid emergence of a well-organized and highly effective worldwide movement in religious environmentalism that in my view will be critically important in the effort to resolve the environmental crisis.

Those who enter this dialogue will discover that the most fundamental scientific truths in contemporary physics and biology are analogous to and entirely compatible with the most profound religions truths in the great religious traditions of the world. They will also learn that recent research in the hard and social sciences has resulted in a radically new understanding of human identity and behavior that has profound moral implications. And they will also take a crash course in environmental science that will provide the conceptual framework needed to understand the causes of environmental problems and how these problems can be resolved.

Environmentally concerned people in the five great religious traditions of the world can enter this dialogue with the assurance that the knowledge we call science cannot in principle be used to dismiss or challenge belief in spiritual reality. But if they challenge the truths of science within its own domain, it will be necessary to either withdraw from the dialogue or engage science on its own terms. Applying metaphysics where there is no metaphysics, or attempting to rewrite or rework scientific truths and/or facts in the effort to prove metaphysical assumptions, merely displays a profound misunderstanding of science and an apparent unwillingness to recognize its successes.

I realize that some readers may now be inclined to believe that the author of this book is a muddle-headed idealist or religious zealot who does not understand the enormous difficulties involved in implementing solutions for environmental problems in the real world. But if these readers will suspend judgment, they will soon discover that this is not the case. For the past two decades I have been actively involved in an effort to implement scientifically viable solutions for environmental problems in the real world and in the process have developed a very realistic understanding of the enormous difficulties involved. And one of the hard lessons learned in developing this understanding should be familiar to profoundly religious people in all of the great religious traditions of the world. The lesson is that faith requires one to be a little crazy in order to be sane. And as the readers of this book should soon discover, this definitely applies to the faith required to resolve the environmental crisis in an insane world in which the business-as-usual approach to resolving this crisis is a recipe for ecological disaster.

There is no space in this short introduction to properly thank all of the friends and colleagues who provided me with invaluable assistance in researching and writing this book. But I did want to take this opportunity to express my very deep gratitude to Professor Peter Brown and to the graduate students in the class we recently team taught at McGill University. And since Peter is among those unsung heroes who has struggled throughout his now very long career to preserve and protect the beauty and wonder of life on this planet and the natural resources that sustain human life, this book is dedicated to him.

CHAPTER 1

Rebirth of the Sacred: Science, Religion, and the New Environmental Ethos

Our house is burning down and we are blind to it. The earth and mankind are in danger and we are responsible. We cannot say that we did not know! Climate change is still reversible. Heavy will be the responsibility of those who refused to fight it.

Jacques Chirac, former president of France

We are Nature, long have we been absent, but now we will return.

Walt Whitman

In *The Dream of Earth*, Thomas Berry makes the following comment about the environmental crisis: "It's all a question of story. We are in trouble now because we do not have a good story. We are in between stories. The old story, the account of how the world came to be and how we fit into it, is no longer effective."[1] The intent in this book is to tell the new story that could greatly enhance the prospect of resolving the environmental crisis. One of the frame tales for this story is science. On the most obvious level, scientific knowledge has gifted us with an understanding of the causes of this crisis and how it can be resolved. What is not so obvious is that this knowledge has also revealed that the old stories about political and economic reality are badly in need of revision.

The old story is imaged on the conventional globes that sit in classrooms, government offices, libraries, and home offices like the one in which I am writing this book. On these globes, boundaries between nation-states are marked with dark lines, and the regions or territories governed by these states are painted different primary colors. The parts (nation-states) are separate and discrete entities, the whole (planet earth) is static, and the sum of the parts constitutes the whole. In the geopolitical reality imaged on these globes, seven billion people live within the borders of sovereign nation-states and construct their identities based on diverse cultural narratives about nationalism, ethnicity, political ideology, and religious beliefs and practices. The only source of political power in this reality is the sovereign nation-state, and these states endlessly compete with one another for the capital and scarce natural resources needed to sustain and grow their national economies.

The new story is imaged in the digital photographs and videos taken by earth-orbiting satellites that environmental scientists use to study the complex web of interactions between human and environmental systems. In these images, there are no divisions between nation-states, the earth is a single dynamic living system, and everything is connected with everything else. Human activities and environmental systems are embedded in and interactive with one another on the local, regional, and global levels, and the state of the whole is a function of the complex web of interactions between all the parts. And as the environmental scientists who study the visual data from these satellites know very well, the scope and scale of global human activities have increased to the point where the capacity of the system of life to support human life is being undermined.

The second frame tale for the new story that could greatly enhance the prospects of resolving the environmental crisis is religion, and this frame will be enlarged to include a dialogue between the truths of science and religion. Those who enter into this dialogue will discover that the most fundamental scientific truths in the new story of science are analogous to and compatible with the most profound religious truths in all of the great religious traditions of the world. The chapters on the new story of science in physics and biology will demonstrate that the part in physical reality we call self is emergent from and embedded in the seamlessly interconnected whole of the cosmos on the micro level and the seamlessly integrated whole of the system of life on the macro level. In this story, any sense that self is separate from world or that mind is separate from nature is an illusion fostered by a lack of understanding of the actual character of physical reality.

The most profound religious truth in the great religious traditions of the world is that human life and consciousness are emergent from and embedded in a single significant whole in spiritual reality and that any sense that self is separate from this whole is an illusion fostered by a lack of understanding of the actual character of this reality. The most profound moral truth in all of these traditions is that the other is oneself and should be treated as one would wish to be treated. Amazingly enough, recent research in neuroscience and the behavioral sciences have revealed that evolution has gifted us with the capacity to act in accordance with this profound moral truth.

This research has not only demonstrated that we have an evolved and innate ability to experience the other as oneself on a precognitive level or in the absence of feedback from higher level cognitive processes associated with language use. It has also revealed that we have an evolved and innate capacity to engage in spontaneous moral behavior that involves great personal risks and sacrifices in the absence of feedback from the higher level cortical processes associated with making conscious moral decisions. Equally significant, recent research in human genetics has shown that all of the seven billion people on this planet are the direct descendents of about two thousand individuals in a small group of hominids living in north central Africa that evolved the capacity to acquire and use fully complex language systems.

The members of the original family of fully modern humans spoke the same language, lived in the same culture, and closely resembled one another in physical terms. After groups of their descendents migrated out of Africa, minor mutations that enhanced the prospects of survival in different ecological niches resulted in changes in physical appearance and language systems, and cultures became increasingly more diverse. But in the new story of science, we are all members of one extended human family and very similar to other members of this family in genetic, cognitive, and behavioral terms.

The scientific truth that frames the new dialogue between the truths of science and religion is that the cause of the environmental crisis and the manner in which it can be resolved are the same. This crisis exists because one species among the millions that have existed on this planet evolved the capacity to acquire and use fully complex language systems and to live storied lives in a linguistically based symbolic universe. And this crisis can be resolved by using this remarkable capacity to create new narratives that can serve as the basis for coordinating human activities in environmentally responsible ways on a planetary scale. But the very inconvenient truth here is that the success of this formidable enterprise will require massive changes in the political and economic narratives and the associated institutional frameworks and processes that now serve as the basis for coordinating global human activities.

The political narrative is premised on the construct of the sovereign nation-state, and the economic narrative is premised on the construct of the invisible hand, or on the belief that the dynamics of market systems legislate over decisions made by economic actors and maintain these systems in relative equilibrium. The chapters on the origins and history of these constructs will reveal that they are products of and deeply embedded in the Western metaphysical tradition and that the legacy of this past is present in metaphysical assumptions about the sources of human identity and the relationship between self and world. This discussion will also demonstrate that these narratives are predicted on unscientific assumptions about the relationship between human and environmental systems that effectively preclude the prospect of implementing scientifically viable solutions for environmental problems.

Since metaphysics will loom large in this discussion, allow me to briefly comment on the meaning of this term. Metaphysics is broadly defined as a "theory of reality in general," and metaphysical assumptions are fundamental aspects of a metaphysical theory. These assumptions are almost invariably perceived as fixed and indelible aspects of reality unless or until they no longer seem in accord with the actual character of experience in the real world. Historically, this has occurred only in situations where massive changes were occurring and societal systems were breaking down. Unfortunately, we have now entered a period in which the escalating impacts of global warming will result in massive changes that could eventually result in the breakdown of societal systems on both the local and global levels. And the little known and rarely discussed reason why this situation exists is that the metaphysical assumptions in the narratives that now serve as the basis for constructing political and economic reality are not in accord with the actual character of physical reality.

A NOT SO MODEST PROPOSAL

In the early 1990s, thirty-four internationally known scientists wrote "An Open Letter to the Religious Community" that appealed to spiritually aware and environmentally concerned people in all religious traditions to create a worldwide religious movement dedicated to the resolution of the environmental crisis.[2] It is not difficult to understand why the scientists who wrote this letter concluded that this movement could be critically important in the effort to resolve the environmental crisis. As Bill McKibben puts it, "Only our religious institutions, among the mainstream organizations of Western, Asian and indigenous societies, can say with real conviction, with any chance of an audience, that there is some point to life beyond accumulation."[3] Or as Roger Gottlieb points out, only religious consciousness can "take us beyond the conventional ego, beyond a frame of mind in which we calculate our interests, struggle for success, seek to control the world to get what we want from it, or unendingly complain about every damn thing we don't have." It is, says Gottlieb, only the experience of "God or Ultimate Reality or Spirit" that makes us feel "less depressed, bored, anxious and selfish and more grateful, joyous and serene" and gives us a sense of "reverence and awe" that we exist in a universe that is "an enthralling mystery."[4]

During the course of this discussion, it should become clear that the massive changes in our political and economic institutions needed to resolve the environmental crisis will not occur in the absence of a well-organized and highly effective worldwide movement in religious environmentalism. The prospects that this could happen are not as unlikely as some may imagine because a worldwide movement in religious environmentalism already exists. This movement consists of alliances between people in different faith traditions, loose networks of spiritually motivated activists, and coalitions with secular environmental organizations. And hundreds of groups of religious environmentalists are now actively engaged in an effort to reduce worldwide emissions of greenhouse gases, eliminate toxic waste dumps, curb losses in biological diversity, and prevent the destruction of wetlands and water sheds.[5]

One of the motive forces behind this movement is a new field of interdisciplinary study known as religion and ecology that emerged in a series of ten conferences held at Harvard Divinity School from 1996 to 1998. During the final stages of this project, representatives of the major religious traditions of the world engaged in an open-ended dialogue with scholars in the fields of science, ethics, economics, education, and public policy about how to resolve the environmental crisis. The papers presented during these conferences appeared in a series of ten groundbreaking books published by Harvard University Press, and work on this project continues at Yale University in the Forum on Religion and Ecology. The editors of these ten books, Mary Evelyn Tucker and John Grim, became the co-directors of the project at Yale, and they managed over the course of a few years to create the largest and most influential international center for the study of religion and ecology.

One fundamental assumption in the work done at the Forum on Religion and Ecology is that the crisis in the global environment is also a religious crisis. Another is that a dialogue between the truths of science and religion that bridges the gaps between religious worldviews and an ecological worldview could greatly enhance the prospects of resolving this crisis. What the readers of this book should soon discover is that our most advanced scientific understanding of physical reality has revealed that there is a new basis for dialogue among these kinds of truths. This dialogue not only provides a compelling and coherent basis for bridging the gaps between religious worldviews and an ecological worldview. It also provides the conceptual framework required to articulate and disseminate an ecological ethos with a profound spiritual dimension that can be embraced by spiritually aware and environmentally concerned people in all religions traditions.

THE SCIENCE OF GLOBAL WARMING

The most menacing environmental problem that must be resolved very soon is global warming. When environmental scientists make presentations at meetings or conferences on the existing and projected environmental impacts of global warming, they describe in jargon-laden language and in emotionally neutral terms what their research has revealed about these impacts. But during informal conversations over a few beers during the evening or late at night, these scientists no longer feel obliged to divorce their scientific heads from their human hearts. On these occasions, they use colorful and often profane language to express their anger toward the politicians and corporate executives who seem incapable of understanding what science has revealed about the existing and projected impacts of global warming.

The scientists involved in these conversations criticize the managers of news media for running stories about drive-by killings, domestic violence, and terrorist activities while saying nothing about the potentially devastating impacts of global warming. And they also tend to express their utter disdain and contempt for the small number of scientists known as global warming skeptics who are well compensated by energy companies and conservative think tanks for misrepresenting and abusing scientific knowledge. But if the conversation goes on long enough and the hour is late, one or more of these scientists will say what the others firmly believe but are reluctant to admit—*the fate of the earth is sealed by the ignorance, lack of compassion, and inexhaustible greed of its human inhabitants, and we will soon find ourselves living on a planet where human existence is little more than a brutal struggle for survival.*

These empirically oriented rational thinkers have reached this dire conclusion because it seems very unlikely in their view that worldwide emissions of greenhouse gases will be reduced to levels where it will be possible to prevent the most disastrous consequences of global warming. But during similar late-night conversations ten years ago, these scientists were not terribly concerned about global warming. The reason why this was the case is that the consensus in the scientific community at the time was that the environmental impacts of global warming would not be a serious problem until about the year 2050.

For the purposes this discussion, there are three important reasons why it is necessary to understand why the scientific consensus about the threats associated with global warming changed very radically over a fairly brief period of time. First, many people have been understandably confused by the fact that many of the same scientists who claimed only a few years ago that global warming is not a serious problem are now claiming that it is a very serious problem indeed. Second, this confusion has been exploited in disinformation campaigns about the science of global warming funded by vested interests in the fossil fuel business and championed by right-wing conservatives who serve these interests in the United States and in other highly industrialized countries. And third, it will not be possible to resolve the environmental crisis unless those of us who care about the human future recognize and come to terms with a very disturbing and quite menacing scientific truth: we are perilously close to the point where global warming could trigger irreversible large-scale changes in the climate system that would threaten the lives of billions of people.

When, during the 1990s, scientists first began to seriously study the complex dynamics involved in the process of global warming, they knew that concentrations of carbon dioxide in the atmosphere prior the beginning of the Industrial Revolution was 275 parts per million (ppm). And they used this data point in the relatively crude climate models that existed at this time to predict the future impacts of global warming. Based on the results of computer simulations using these models, scientists concluded that global warming would not be a serious threat to human populations until concentrations of carbon dioxide in the atmosphere reached 550 ppm about the year 2050.

The usual explanation why this prediction was grossly inaccurate is that scientists did not have sufficient observational data to accurately assess the existing impacts of global warming and computer systems that could realistically model future impacts. But what those who offer this explanation fail to mention is that there was no compelling scientific evidence to justify the claim that the impacts of global warming on environmental systems would be relatively benign until concentrations of carbon dioxide in the atmosphere reached 550 ppm. The only justification for making this claim was that 550 ppm was twice the concentration of carbon dioxide in the atmosphere prior to the Industrial Revolution. But in fairness to the scientists, this claim seemed perfectly reasonable at the time because previous research had revealed that marginal increases in the levels of carbon dioxide during the Industrial Revolution had minimal impacts on environmental systems. Based on the assumption that future impacts would be similar to those that occurred in the past, the scientists concluded that additional increases in the levels of carbon dioxide would have minimal impacts and that any changes in environmental systems would be glacially slow.

Since the consensus in the scientific community during the 1990s was that global warming would not be a serious problem until about the year 2050, it seemed that there was enough time to deal with this problem. This explains why virtually all political leaders and business planners assumed that the problem could be resolved with incremental reductions in worldwide emissions of carbon dioxide over an extended period of time. And it also explains why this gradualist and incremental approach was embraced by scientists associated with the Intergovernmental Panel on Climate Change (IPPC) and by environmentalists working for mainstream nongovernmental organizations (NGOs).

Meanwhile, data from satellites and ground observation systems indicated that the large-scale impacts of global warming were occurring much sooner and far faster than the existing computer models predicted. Scientists were perplexed by this alarming development and could not initially explain why these impacts were occurring. But they eventually realized that the predictions based on the existing models were grossly inaccurate because they did not take into account the fact that the global environment is a nonlinear system. In these systems, there typically are large numbers of initial conditions, many degrees of freedom, or directions in which the systems can develop or evolve, and the future of these systems can be predicated only within a range of probabilities.

Until recently, scientists avoided the study of nonlinear systems due to their seemingly chaotic nature and the fact that nonlinear equations are usually too difficult to solve. This situation changed dramatically with the availability of powerful computers that could compute the large number of values that satisfy nonlinear equations and provide solutions represented graphically as a curve or set of curves. Scientists now use nonlinear systems theory to study the nonlinear dynamics of climate change, and this theory is also known as systems dynamics and complex dynamics. But nonlinear systems theory is not a theory as that term is normally used in science. It is a collection of mathematical techniques that applies to a broad range of phenomena, and the same is true for two important branches of this theory—chaos theory and the theory of fractals.

After scientists began to study the nonlinear dynamics of the environmental impacts of global warming with more sophisticated climate models and computer systems, they soon realized that initial impacts were greatly amplified by self-reinforcing positive feedback loops. A feedback loop is essentially a series of causally connected elements in which an initial element causes effects on other elements that propagate around a loop and thereby feed back into the original element. In other words, an original input affects the last output in repeated cycles, and each input is modified or changed by the previous cycle. A feedback loop is called positive if recurrent inputs amplify the initial change and move the system in a particular direction and negative if these inputs counteract or suppress the initial change and move the system back toward its original state. In mathematical terms, the loops are represented in a special kind of nonlinear process called iteration in which a function operates repeatedly on itself.

THE BAD NEWS ABOUT GLOBAL WARMING

When satellite and ground observational systems indicated in 2007 that marginal increases in average temperature in the Arctic were resulting in the melting of increasingly larger amounts of ice, scientists simulated the nonlinear dynamics involved with the use of more sophisticated climate models and computer systems. And these simulations revealed that there were self-reinforcing positive feedback loops between marginal increases in average temperature, the melting of increasing larger amounts of ice, the replacement of the ice by sea water that absorbs vastly more sunlight than the ice, and additional increases in average temperature.[6]

Observational data also indicated in 2007 that marginal increases in average temperature in the Arctic were resulting in the release of enormous amounts of the very potent greenhouse gas methane from the frozen tundra and the frozen clathrates at the bottom of the ocean. After scientists simulated the nonlinear dynamics involved, they soon realize why this very alarming development was occurring. These simulations revealed that there are self-reinforcing positive feedback loops between small increases in average temperatures in the Arctic, the melting of more tundra and clathrates, the release of increasingly larger quantities of methane into the atmosphere, and additional increases in average temperature.[7]

Equally ominous, data from ground and satellite observational systems showed in 2007 that polar ice sheets were melting fifty years sooner than previous computer models had predicted[8] and that the expansion of the tropics during the past two decades was greater than these models predicted would occur by the end of the twentieth century.[9] Once again, computer simulations of the nonlinear dynamics involved revealed that the culprit was self-reinforcing positive feedback loops. During this same year, scientists also demonstrated that these feedback loops were responsible for the evaporation of more water from tropical forests and wetlands, the drying out and decay of vegetation in these regions, and the release of increasingly larger amounts of carbon dioxide.

During the 1990s, it was widely assumed in the scientific community that if worldwide emissions of carbon dioxide dramatically decreased, this would result in a sustainable global environment and the problem of global warming would be resolved. Unfortunately, recent scientific research has demonstrated that this assumption is false. One reason why this is the case is that the life span of carbon dioxide in the atmosphere is more than a century and that there is already enough greenhouse gas in the atmosphere to raise global temperature by two degrees Centigrade.[10] Another is that the self-reinforcing positive feedback loops that will result from this unavoidable increase in average temperature will have massive impacts on environmental systems that will not be reversible for a very long time.

For example, a study published in January 2009 concluded that changes in "earth temperature, rainfall, and sea levels" caused by global warming would be irreversible "for more than a thousand years after carbon dioxide emissions are completely stopped."[11] And a study published in July 2009 showed that "coral reefs cease to be viable in the long run"[12] when concentrations of carbon dioxide in the atmosphere reach 360 ppm. Even more disturbing, scientists demonstrated in a paper published online in October 2009 that when concentrations of carbon dioxide in the atmosphere exceeded 350 ppm twenty million years ago, sea levels increased by a hundred meters or more and temperatures increased by about ten degrees Celsius.[13]

THE NEW TERMS OF HUMAN SURVIVAL

The research that demonstrated that allowing concentrations of carbon dioxide to reach 550 ppm would have disastrous consequences was done by climate scientist James Hansen and his team at NASA. When the results were published in 2008, one of the disturbing conclusions was that the large-scale impacts of global warming would occur much sooner and faster than previous computer models had predicted. But the prediction that sent shock waves though the scientific community was that if we fail to reduce concentrations of carbon dioxide in the atmosphere from the present 400 ppm to 350 ppm over the next twenty years, there is a high probability that global warming will trigger irreversible large-scale changes in the climate system and environmental subsystems.[14] As the author of the lead article in the scientific journal *Nature* put it in September 2009, concentrations of carbon dioxide in the atmosphere in excess of 350 ppm would "threaten the ecological life support systems" and "severely challenge the viability of contemporary human societies."[15]

If we fail to prevent global warming from triggering irreversible large-scale changes in the climate system, the consensus in the scientific community is that we will soon find ourselves living in a very different world. In this world, there would be rapid melting of the glaciers in the Himalayan–Tibetan plateau, the Andes and Rocky mountains, more gradual melting of the ice sheets in Greenland and west Antarctica, the dieback and eventual disappearance of the Amazon rainforest, and more persistent and higher latitude El Nino conditions. Dramatic increases in the intensity and frequency of floods and droughts would cause crop failures and declines in agricultural production in virtually every country. People living in developed countries like the United States would experience food and water shortages, and those living in developing and underdeveloped countries, particularly in South Africa and central and south Asia, may not have sufficient food and water to survive.

The melting of the ice sheets in Antarctica and Greenland would cause sea levels to rise by one to two meters, and densely populated low-lying regions in the Indian subcontinent and elsewhere would become uninhabitable. Severe flooding would occur in Miami, New York, Newark, New Orleans, Boston, Washington, Philadelphia, Tampa/St. Petersburg, and San Francisco. The cities outside of North America that also would be inundated by rising water levels include Osaka/Kobe, Tokyo, Rotterdam, Amsterdam, and Nagoya. The gradual disappearance of the mountain glaciers in the Himalayas would eventually deprive over a fourth of the global human population of their primary source of fresh water. In the most severely affected countries in south and central Asia, mass migrations of starving and desperate people into neighboring countries would probably result in cross-border conflicts that could easily escalate into full-scale wars. In one of the regions where this massive migration is likely to occur, the full-scale war could be between the nuclear-armed countries of India and Pakistan.[16]

The irreversible large-scale changes in the climate system and environmental subsystems would also massively disrupt production and distribution systems in the global market system. These disruptions would result in huge financial losses by business owners and investors, dramatic declines in savings and capital investments, rapid increases in the prices of scarce commodities and foodstuffs, runaway inflation, and massive unemployment. Banking systems and stock markets in underdeveloped and developing countries would collapse, and those in developed counties would suffer catastrophic losses. In contrast with previous financial crises, productivity, profit margins, and returns on investment in stocks, bonds and other financial instruments would continue to decline. And since there would be no available capital for investment spending, the global market system would eventually collapse.

THE POLITICS OF GLOBAL WARMING

This dramatic change in the scientific consensus about the threats of global warming has been massively exploited in the disinformation campaign of the climate change contrarians. The strategy used in this campaign was articulated by Republican pollster Frank Lutz in a 2003 memo to the Bush administration: "The scientific debate is closing against us but not yet closed. There is still a window of opportunity to challenge the science. Voters believe that there is no consensus about global warming within the scientific community. Should the public come to believe the scientific issues are settled, their views will change accordingly. Therefore, you need to make the lack of scientific certainty a primary issue in the debate."[17]

The campaign of the climate change contrarians gained considerable momentum after emails written by environmental scientists associated with the Intergovernmental Panel on Climate Change (IPCC) at the University of East Anglia were hacked and widely distributed over the internet in November 2009. Some of these emails in this private correspondence contained unkind words about research done by scientists in the employ of the petroleum industry, and others contained statements that could be interpreted out of context as suppressing this research. But in the view of the contrarians, these comments constituted sufficient evidence to charge all IPCC scientists with everything from professional misconduct to engaging in a conspiracy to suppress scientific research that was not in accord with their ideological and political agendas.

This campaign gained even more momentum after some contrarians claimed that IPPC scientists made two scientifically inaccurate predictions about the environmental impacts of global warming in their 2007 report. What those who made this claim failed to realize is that one of these predictions, that the Himalayan glaciers would disappear by 2035, was made in a 738-page working paper and did not appear in the 2007 report. The other prediction about crop failures in North Africa did appear in the background information section of the three-thousand-page report but had no bearing whatsoever on the conclusions drawn. Nevertheless, the lawyers for the prosecution in the campaign of the climate change contrarians accused the IPCC scientists of committing fraud and coined the term *climategate* to refer to an alleged conspiracy to cover up or suppress the truth about global warming.

The climate change contrarians could be dismissed as laughably absurd if there were not in the process of achieving their objectives. A poll taken by the Pew Research Center indicated that the percentage of Americans who believe there is solid evidence of global warming declined from 71 percent in April 2008 to 57 percent in October 2009. This poll also revealed that only 35 percent of Americans believe that global warming is a "very serious problem," only 36 percent think it is caused by human activities, and only 30 percent said it should be a top priority for President Barack Obama and members of Congress.[18] Recent polls also suggest that the campaign of the climate changes contrarians in Europe has been quite successful. BBC surveys found that the percentage of Britons who believe that "climate change is happening and is now established as largely man-made" declined from 41 percent in November 2009 to 26 percent in May 2010. And a similar poll by the German magazine *Der Spiegel* indicated that in May 2010, 42 percent of Germans believed that global warming is a serious problem, down from 62 percent four years ago.[19]

Numerous efforts have been made to explain why the campaign of the climate change contrarians has been such a stunning success. Some have argued that the vast majority of people in this country and in other highly industrialized Western counties are scientifically illiterate and easily duped into believing the lies told by the contrarians. Others attribute the success of this campaign to the fact that the managers of news media fail to recognize or choose to ignore the distinction between the science and politics of global warming and violate the public trust by providing the contrarians with a forum for disseminating their lies. Another standard explanation for this success is that the contrarians receive large financial support from wealthy individuals and large corporations and operate in concert with conservative think tanks and lobbying organizations. And the psychological explanations typically attribute this success to the human tendency to deny the existence of global warming when confronted with anxiety-inducing scientific descriptions of its extant and projected impacts.

But what those who offer these explanations have apparently failed to realize is that the most pervasive theme in the campaign of the climate change contrarians is not that scientists are engaged in a vast conspiracy to delude the public into believing that global warming is a serious problem caused by human activities. It is that any systematic attempt by government to control and regulate emissions of greenhouse gases would undermine the autonomy and integrity of free market systems, retard the growth and expansion of these systems, and threaten individual rights and freedoms. The reasons why this is the most pervasive theme in the campaign of the climate change contrarians will become apparent in the chapters on the origins and history of the construct of the invisible hand in mainstream economic theory. This discussion will not only reveal that the climate change contrarians and those who support this campaign are true believers in what will be termed here the *religion of the market*. It will also demonstrate that this religion has a well-defined set of beliefs and practices and an elaborate theological framework.

The next chapter contains a brief account of what the new story of science has to say about how members of one species among the millions that have existed on this planet became fully conscious and self-aware beings in the vast cosmos. This remarkable development occurred because the highly improbable and utterly unique evolutionary pathway of our species resulted in the ability to acquire and use fully complex language systems. Scientific research on the preadaptive changes that made this development possible has also resulted in a new understanding of human nature and the sources of human indentify that could be critically important in the effort to resolve the environmental crisis. Equally important, this new story of science provides a coherent basis for understanding how our species became the greatest destroyer of life on this planet since a meteor put in an end to age of the dinosaurs sixty-five million years ago.

CHAPTER 2

The New Story of Humanity: Brains, Minds, and the Environmental Crisis

What a piece of work is man, how noble in reason, how infinite in faculties, in form and moving how express and admirable, in action like an angel, in apprehension how like a god, the beauty of the world, the paragon of animals.

William Shakespeare

Man in his arrogance thinks himself a great work, worthy the imposition of a deity. More humbly I believe true to consider him created from animals.

Charles Darwin

While sitting in a window seat during a flight from San Francisco to Washington, D.C. about twenty years ago, I had an experience that changed the course of my life. On the ground below, vast numbers of trucks and mile-long strings of railroad cars were moving along extensive networks of highways and tracks that threaded out in all directions, like a circulator system in some giant organism. Products from factories and farms were flowing through these arteries toward distant cities and coastal ports, and raw materials were flowing in the other direction to processing and manufacturing plants.

In my mind's eye, the web-like connections between electric power plants, transformers, cables, lines, phones, radios, televisions, and computers resembled the spine and branches of a central nervous system, and the centers of production, distribution, and exchange seemed like tissues and organs. Next, I enlarged this frame to include all centers of production, distribution, and exchange and all connections between them within the global economy. This conjured up the image of a superorganism feeding off the living system of the planet and extending its bodily organization and functions into every ecological niche. I realized, of course, that the global economic system is not an organism. It is a vast network of technological products and processes that members of our species created in an effort to enhance their material well-being. But this system does in ecological terms feed off the system of life on this planet and extend its organization into every ecological niche.

After my plane landed at Dulles International Airport, I asked a simple question that required years of research to adequately answer. How did members of one species among the millions of species that have existed on this planet manage to increase their numbers and the scope and scale of their activities to the point where the capacity of the system of life on an entire planet to support their existence is being undermined? The answer is that our species, fully modern humans, evolved against all odds the capacity to acquire and use fully complex language systems.

The incremental changes in the brains and bodies of our hominid ancestors that culminated in the ability to acquire and use fully complex language systems probably occurred over a period of about two million years. But there is now a large and growing consensus that this process was not complete until about sixty-four thousand years ago, when a small group of hominids living in present-day Ethiopia, Kenya, and Tanzania evolved the capacity to use a fully complex language system with grammar and syntax. Even more remarkable, recent studies in human genetics have revealed that the seven billion people living on this planet today are all direct descendents of about two thousand individuals in this small lineage of hominids.[1]

The ability to coordinate experience in a linguistically based symbolic universe made fully modern humans increasingly more immune to the usual dynamics of evolution that control the relative size of populations in particular ecological niches. The reason why this is the case is apparent in the artifacts found in the fossil remains. When we examine these remains prior to sixty-four thousand years ago, the stone tools are primitive, display little innovation, and are similar to those used by the Neanderthals. There were no unequivocal compound tools, such as a wooden handle with an axe-like blade, and no variations in tool-making in different geographical locations.

But artifacts found in the fossil remains of the fully modern humans who lived in France and Spain forty thousand years ago grandly testify to their creativity and intelligence. Compound tools, standardized bone and antler tools, and tools that fall into distinct categories or functions, such as mortars and pestles, needles, rope, and fishhooks, appear in these fossil remains. Also found in these remains are weapons designed to kill large animals at a distance, such as darts, barbed harpoons, bows and arrows, and spear throwers. And some of these artifacts, such as rock paintings, necklaces, pendants, fired-clay ceramic sculptures, flutes, and rattles, clearly indicate the existence of profound aesthetic preoccupations and religious impulses.

Why did complex human societies suddenly appear after a two-million-year period in which our hominid ancestors lived in extremely primitive conditions that remained virtually static? The most reasonable explanation is that a series of mutations that occurred about sixty-four thousand years ago in the brains of the hominid ancestors living in northern Africa resulted in a much higher level of integration between the neuronal processes and pathways associated with the use of fully complex language systems. Some experts argue that these mutations took place over something like one hundred generations during a period in which selection pressures due to climatic stress were very high and mutations that enhanced language abilities provided a distinct survival advantage.[2] And this hypothesis has been reinforced by recent studies in linguistics that strongly suggest that all extant human languages have a common origin in a single language system that existed about fifty thousand years ago.[3]

But even if this hypothesis is not correct, the result was utterly amazing. A lineage of hominids that had previously seemed unimaginative, only marginally intelligent, and destined for extinction became fully conscious and self-aware individuals who lived storied lives in a linguistically based symbolic universe. In this universe, experience could be represented and organized in themes and narratives, and the terms of survival could be altered and manipulated with complex social behavior and ideas externalized as artifacts. Our hominid ancestors may have been very similar to other surviving species of hominids prior to the point where they evolved the capacity to acquire and use fully complex language systems. But after this momentous event occurred, the difference between our species and other life-forms, including our primate cousins, became a yawning chasm.

Until recently, very little was known about the myriad number of mutations that culminated in the capacity to acquire and use fully complex language systems. But this situation changed dramatically over the past two decades due to advances in neuroscience made possible by computer-based brain imaging systems, such as positron emission tomography (PET) and functional magnetic resonance imaging (fMRI). These systems allow researchers to observe which areas in the brains of conscious subjects are active while performing specific cognitive tasks.

Research based on these technologies has revealed that language processing is staggeringly complex and places incredible demands on memory and learning. Language functions extend into all major lobes of the neocortex—auditory input is associated with the temporal area; tactile, with the parietal area; and attention, working memory, and planning, with the frontal cortex of the left or dominant hemisphere. The left prefrontal region in people with normal hemispheric dominance is involved with verb-and-noun production and in the retrieval of words representing action. Broca's area, adjacent to the mouth–tongue region of the motor cortex, plays a major role in vocalization for word formation, and Wernicke's area, adjacent to the primary auditory cortex, manages sound analysis in the sequencing of words (see figure 2.1).

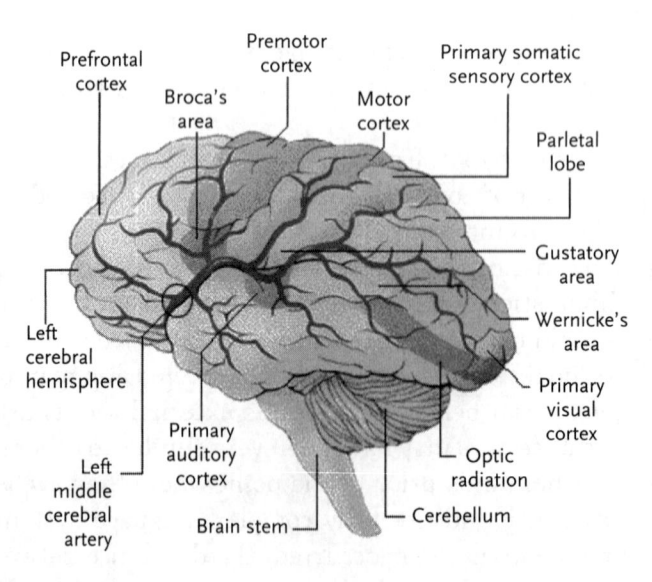

Figure 2.1
The Human Brain

We now know that lower brain regions also evolved in ways that contributed to ability to acquire and use fully complex language systems. For example, the cerebellum was previously thought to be exclusively involved with automatic or preprogrammed movements, such as throwing a ball, jumping over a high hurdle, or playing well-practiced notes on a musical instrument. But imaging studies in neuroscience indicate that the cerebellum is activated during speaking and activated most when the subject is making difficult word associations. The cerebellum also plays a role in association by providing access to automatic word-sequences and by augmenting rapid shifts in attention. Similarly, the midbrain and brain stem, which sit on top of the spinal cord, coordinate input and output systems in the head and play a crucial role in communication functions involving vocalization.

THE AMAZING GIFT OF LANGUAGE

Anyone who doubts that evolution has gifted the members of our species with the innate capacity to acquire and use fully complex language systems with little or no formal training should spend some time observing language learning in infants and children. A four-day-old infant can discriminate between the voice of the mother and another woman the same age, between a natural flow of speech and words spoken in isolated sequence, and between the language spoken by the mother and another language.[4] At three months, the larynx descends in the throat, opening up a space behind the tongue, and this allows the infant to produce basic sounds, or phonemes, used by others in the linguistic environment by moving the tongue backward and forward.

An eight-to-ten-months-old child will normally babble, or utter syllables in an apparently meaningless fashion, and then begin to group phonemes into the word symbols used by its caretakers. Babbling is one of the early signs that vocal motor output is being activated via neuronal pathways that are different from those involved in more basic forms of vocalizations, such as crying. This output is controlled in part by the motor cortex, and the onset and maturation of the babbling process corresponds with the growth of cortical output pathways.

Prior to eighteen months, the inner narrator that will later be identified as the conscious self does not exist, and children at this stage of development cannot identify themselves in a mirror as a self different from other selves. After about eighteen months, a normal child acquires a new word about every two hours and will normally continue to expand its vocabulary at this rate through adolescence. Between eighteen and twenty-four months, children begin to separate the contents of other people's minds from their own beliefs and are capable of pretending that something is true when it is not. From the late twos and the mid-threes, children begin to speak fluent grammatical sentences so rapidly that researchers have not been able to chart the exact sequence. And it is only after children are capable of speaking these sentences that they become aware of an inner self that self-consciously focuses attention on the outside world and has thoughts that follow the logic of its own internal experience.

A normal infant can acquire any language system with ease before the age of four, and this occurs even in cultures where mothers rarely speak to their prelinguistic children. At four, children are able to attribute to others beliefs they know to be false and are engaged thereafter in the arduous business of separating truths from untruths.[5] At five or six, children possess a vocabulary of several thousand words, and word learning continues at a prodigious rate throughout adolescence and into early adulthood. It has been estimated that an average American high school student knows about forty-five thousand words, about three times the number used by Shakespeare,[6] and that the vocabulary of the average college graduate consists of about sixty-thousand words.[7]

What is most remarkable about human language systems is that they are enormously pliable and capable of generating an infinite variety of meanings. One reason why this is the case is that grammar, or the neuronal patterns, pathways, and feedback loops associated with grammatical constructions, allows a finite number of discrete elements to be combined into larger structures with properties that cannot be reduced to or explained in terms of the parts. Another is that the thoughts constructed in language are combinatorial (parts combine) and recursive (parts can be embedded in other parts), and this allows a great deal of knowledge to be generated and explored within a finite variety of structures. This explains why the number of sentences a normal person is capable of uttering is astronomical and increases exponentially as vocabularies enlarge. It also explains why, on average, ten different words could be inserted at any random point in a sentence to complete the sentence in a grammatical and meaningful way.

A NEW VIEW OF HUMAN NATURE

It now appears that one of the preadaptations that contributed to the evolution of the brain regions and neuronal pathways that allowed our ancestors to acquire and use fully complex language systems is the mirror neuron system. Mirror neurons are concentrated in the premotor cortex, the posterior parietal lobe, the superior temporal sulcus, and the insula, and they fire when we perform a particular action or watch this action performed by others. In addition to simulating the action or behavior of others in the brain of the observer, the mirror neuron system also allows the observer to feel the emotions and grasp the intent associated with the action or behavior. Equally interesting, this system operates with little or no feedback from higher level cognitive processes and conscious reflective behavior. As the neuroscientist Giacomo Rizzolatti has put it, "Mirror neurons allow us to grasp the minds of others not through conceptual reasoning but through direct simulation. By feeling, not by thinking."[8]

Given that the mirror neuron system provides a basis for imitation learning and simulating the behavior of others on a preverbal or nonverbal level, it probably played a critically important role in the evolution of a gesture performance communication system. And a great deal of evidence suggests that this system served as the neuronal substrate for the evolution of the specialized brain regions and neuronal pathways associated with the use of fully complex language systems. For example, the neuronal processes involved in learning syntax appear to have originated in pointing gestures, which children use in the early stages of language learning. Mirror neurons are also highly concentrated in Broca's area, a brain region associated with both language use and the imitation of the behavior of others.

It has been known for some time, as psychologist Seymour Epstein puts it, that "people apprehend reality in two fundamentally different ways, one variously labeled intuitive, automatic, natural, non-verbal, narrative, experiential, and the other analytical, deliberative, verbal, and rational."[9] But what was not known until quite recently is that the brain regions and neuronal processes associated with the first of these fundamental ways in which we apprehend reality evolved prior to those associated with the use of fully complex language systems. Recent research in neuroscience and the behavioral sciences has shown that this much older nonverbal system encodes reality in images and metaphors, makes connections through the process of association, and mediates behavior based on feelings from past experiences. This research has also revealed that there are extensive feedbacks from this system to the limbic system and that these feedbacks result in powerful feelings of empathy, sympathy, and compassion. The more recently evolved verbal system operates very differently. This system encodes reality in abstract symbols, makes connections by logical analysis, mediates behavior by conscious appraisal of events, and results in deliberative behavior. And since the feedbacks from this system to the limbic system are minimal, the process of making conscious moral decisions tends not to be motivated by powerful feelings of empathy, sympathy, and compassion.[10]

The nonverbal system associated with spontaneous moral behavior probably enhanced the prospects of survival in small hunter–gatherer tribes in which the lives of individuals were often threatened by present, visible, and immediate dangers. The legacy of this evolutionary past is apparent in research in the behavioral sciences on the spontaneous moral behavior of surviving members of the extended human family. This research has shown that people are far more willing to provide aid that could relieve the suffering of identified individuals with names and faces than for unidentified individuals or groups of such individuals.[11]

Participants in one of these studies were given the opportunity to contribute $5 of their earnings to the Save the Children Foundation to provide relief for victims of a severe food crisis in Africa and Ethiopia. They were then asked to indicate how much money they would be willing to donate to provide relief for people in the following categories: (1) an identifiable victim with a name and face, (2) numerous victims described in numerical terms, and (3) an identifiable victim described in these terms. The identifiable victim was a small boy named Rokia, and his face could be seen in a photograph. Those involved in this study donated much more money to provide aid for Rokia than for unidentified victims or for numerous victims whose suffering was described in numerical terms. But when the photograph of Rokia was accompanied by a description of his plight in these terms, the amount of aid that the participants in this experiment were willing to provide declined significantly.[12]

This research seems to provide a basis for answering a question that has perplexed moral philosophers and theologians for a very long time: Why are good people who are quite willing to make personal sacrifices to preserve and protect the lives of individuals unwilling to make these sacrifices to preserve and protect the lives of large numbers of people? The first part of the two-part answer is that the spontaneous moral behavior that results in a willingness to make personal sacrifices for identified individuals is associated with neuronal processes in the nonverbal system that operates on the unconscious level. The second part is that the process of making decisions about whether to make personal sacrifices for large numbers of unidentified individuals is associated with neuronal processes in the verbal system that operate on the conscious level.

When we encounter the suffering of identified individuals, the unconscious nonverbal system is activated, and positive feedback loops between this system and the source of emotions in the limbic system result in feelings of empathy, sympathy, and compassion. But when we are confronted with numerical descriptions of the suffering of large numbers of unidentified individuals, this information is processed in the conscious verbal system. And since this processing occurs in the absence of massive feedbacks to the limbic system, the decisions to make personal sacrifices that could relieve this suffering tend not to be motivated by feelings of empathy, sympathy, and understanding.

The stark differences between spontaneous moral behavior associated with the unconscious nonverbal system and making decisions that could have moral consequences associated with the conscious verbal system are apparent in the following comment by the late Mother Teresa of Calcutta: "If I look at the mass I will never act. If I look at the one, I will." Another comment made by novelist Annie Dillard nicely illustrates the difficulties involved in reconciling these differences: "There are 1,198,500,000 people alive in China. To get a feel for what this means, simply take yourself—in all your singularity, importance, complexity and love—and multiply by 1,198,500,000. See? Nothing to it."[13]

There is now a large and growing consensus in both the hard and behavioral sciences that the human capacity to engage in spontaneous moral behavior is a product of evolution and is innate. And research in the behavioral sciences strongly suggests that the moral concepts and emotions associated with this behavior are universal in spite of the differences in standards for ethical behavior in diverse cultural contexts. For example, anthropologist Donald Brown has compiled an impressively long list of these universal moral concepts and emotions, which includes distinctions between right and wrong; empathy; fairness; rights and obligations; prohibitions against murder, rape, and other forms of violence; shame; taboos; and sanctions for wrongs against the community.[14]

Studies done by anthropologists have also shown that people living in existing hunter–gatherer tribes display a strong belief in fairness and reciprocity, a great capacity for empathy and impulse control, and a pronounced willingness to work cooperatively for the good of the entire community. And numerous studies done on both children and adults living in highly industrialized Western countries have revealed that a violation of the expectation that others will display a sense of fairness evokes feedbacks from the limbic system associated with outrage and indignation.[15]

A BRIEF EARLY HISTORY OF THE HUMAN FAMILY

The members of the original family of fully modern humans spoke the same language, lived in the same culture, and closely resembled one another in physical terms. The first migration out of East Africa about fifty thousand years ago was apparently north along the Nile Valley and across the Sinai Peninsula into the Middle East, the Near East, southeastern Europe, and southwestern Europe. The ancestors who crossed over the Asian mainland into Australia and New Guinea between forty and thirty thousand years ago probably made use of some form of watercraft. Other members of this extended family moved along the coastlines of India and southeastern Asia, and some evidence suggests that about fourteen thousand years ago their descendents crossed the land bridge that joined Siberia and Alaska into North and South America.[16]

As small groups of fully modern humans migrated over the course of many generations to more distant regions, minor mutations occurred that enhanced survival in disparate climatic conditions, and language systems and cultures became increasingly diverse. For those living in equatorial regions where ultraviolet rays from the sun are intense, dark pigmentation helped to prevent skin cancer and severe sunburn. But in regions where these rays were less intense, dark skin was a liability because it did not allow enough ultraviolet light to penetrate the skin and synthesize a sufficient amount of vitamin D to prevent the painful and disfiguring disease of rickets.

In this situation, minor mutations in the genes of people who lived in these regions resulted in lighter skin pigmentation.[17]

Other minor mutations in the genes of members of our extended human family who eked out their existence for many generations in the Siberian snowfield resulted in the epicanthic folds over their eyes that enhanced vision in cold winds and reduced the glare of the sun. Similarly, the Pygmies of equatorial Africa have smaller bodies because mutations that reduced body mass were conducive to survival in sweltering rain forests, and the Africans and aboriginal Australians have frizzier hair and broader noses because mutations resulting in these features enhanced the prospects of survival in hot climates.

During the course of these migrations, fully modern humans invented new words and developed alternate ways of inflecting words and arranging sequences of words in grammatical and syntactical terms. If two previously isolated groups came into contact with one another after only a few generations, the members of one group would have been able to understand those in the other without much difficulty. But this would not have been the case if they were separated for about a thousand years because the languages spoken by each would have changed so much that there would be no basis for meaningful communication or mutual comprehension.[18]

The story about how our species became the greatest destroyer of life since a meteor struck the earth and put an end to the age the dinosaurs begins after a group of fully modern humans settled some twelve thousand years ago in a region in the Middle East known as the Fertile Crescent. Bounded by the Tigris and Euphrates rivers, the Fertile Crescent featured a wide range of altitudes and a great diversity of climactic conditions. The mild, wet winters and hot dry summers favored the evolution of a variety of plants with large seeds that could survive the dry season and readily sprout when the rainy season began. Thirty-six of the fifty-six species of wild grasses on this planet that are suitable for domestication because of their large seeds grew in concentrated abundance in the Fertile Crescent. Other species existed in regions with similar climates, but they were far fewer in number, scattered over larger territories, and less suitable for domestication.

The ecology of the Fertile Crescent, particularly the abundance of wild grasses with large seeds, was conducive also to the evolution of four large group-living, grazing mammals—the goat, sheep, the pig, and the cow. These animals were passive, social, and amenable to human manipulation and control. Over a period of several thousand years, the fully modern humans who settled in this region raised these animals in captivity and modified their appearance and behavior by breeding them and taking control over their food supply. Other large mammals existed in areas with similar climates, such as California, Chile, southwestern Australia, and South Africa, but none was as suitable for domestication.

In the initial phase, the settlers in the Fertile Crescent collected large quantities of naturally growing, ripe wild cereals and stored the seeds for use later in the year. Eventually, they took the seeds from the hillsides where rain was unpredictable and planted them in damp river bottoms where the growth of the plants was less dependent on water supplied by intermittent rain. Over a period of several thousand years, an agricultural society emerged in which the plants were systematically grown in fields and the animals were used for fertilizer, milk, wool, plowing, and transport. This eventually resulted in a system for intensive food production comprised of three cereals that were the main source of carbohydrates, four pulses (legumes) that provided some protein, and four domestic animals that were the principal sources of protein. The first societies that were entirely dependent on crops and domesticated animals for their survival appeared in the Fertile Crescent about eight thousand years ago.

HUMAN TRANSFORMATION OF THE BIOSPHERE

The creation of these societies was the first phase of a process that eventually would transform the entire biosphere, an approximately twenty-three-mile sphere that extends from the depths of the ocean to the top of the troposphere. The mutual interactions of organisms within this sphere produce and remove gases, ions, metals, and organic compounds, mediate the growth and metabolism of organisms, and modulate temperature, alkalinity, and atmospheric compositions. If earth could be reduced to the size of a basketball, the outer layer of the biosphere would be thinner than the finest paper.

In the initial stage of the human transformation of the biosphere, the wild plants in the Fertile Crescent initially cultivated as crops (wheat, barley, and peas) were already edible, existed in abundance in the wild, and could be easily sown and harvested within a few months. These plants were self-pollinating, produced seeds that could be stored for later consumption, and all that was required to make them more suitable as crops was to plant the seeds of mutated plants that had more desirable characteristics. For example, the seeds of mutated wheat plants that had nonshattering stalks were used to grow more wheat, and these mutated plants became the ancestors of the domesticated plants that now produce all crops of wheat.

During the second stage, which began about 4000 b.c., fruit and nut trees in the Fertile Crescent that could be grown by simply planting cuttings or seeds were domesticated to produce olives, figs, dates, and pomegranates. Given that these crops did not yield food until at least three years after they were planted and did not reach full productive capacity for about a decade, they could be grown only by people in settled villages who were capable of long-range planning. The third stage involved the domestication of trees that produced apples, pears, plums, and cherries, which could not be grown from cuttings and tended to yield highly variable and often worthless fruit when grown from seeds.

We now know that only a few thousand of roughly two hundred thousand wild plants are consumed by humans, and only a few hundred of these are grown as domesticated crops. We also know that most of these crops produce food that merely supplements the human diet and that over eighty percent of the annual tonnage of all crops comes from a mere twelve species—the cereals wheat, corn, rice, barley, and sorghum; the pulse soybean; the tubers potato, manioc, and sweet potato; the sugar-producing sugarcane and sugar beet; and the fruit banana. And only three of these species, the cereals wheat, rice, and maize, account for more than half of the human intake from all plants. Virtually all of these crops were cultivated by Roman times, and not one new major food plant has been domesticated in modern times.[19]

The domesticated plants and animals in the agricultural system in the Fertile Crescent originally constituted about 0.1 percent of the total biomass of the planet. But as this system expanded, these plants and animals increased in number well beyond the limit that would have been possible during the natural course of evolution. The dramatic result is that the plants and animals in the present global agricultural system, providing that we exclude the invertebrates, account for a staggering 90 percent of the terrestrial biomass of earth.

During the process of transforming the biosphere, our species destroyed about 50 percent of the wetlands, 33 percent of the mangroves, 60 percent of the hardwood and mixed forests, 30 percent of the conifer forests, 45 percent of the tropical rain forests, and 70 percent of the tropical dry forests. This process resulted in the loss of 90 percent of the large predator fish, overfishing in 75 percent of marine fisheries, and hundreds of dead zones in the oceans caused by runoffs from factories, farms, and sewage treatment plants. Human activities have also increased the rate at which species are becoming extinct to a thousand times faster than normal.[20] We now consume or destroy about forty percent of the photosynthetic output of the biosphere every year and have appropriated virtually all of the land suitable for agricultural production. And if all of the surviving members of the extended human family consumed the same amount of environmental resources as the average citizen of the United States, two additional planet earths would be needed to supply these resources.[21]

HUMAN POPULATION GROWTH AND THE ENERGY REGIME OF OIL

The population explosion between 8,000 b.c. and 3,000 b.c. from about five million to one hundred million was fueled by more systematic exploitation of the energy contained in the seed plants and domesticated large mammals in the Fertile Crescent. After about 4,000 b.c., the increase in the number of people living at higher population densities could only be sustained by new technologies that allowed more energy to be stored and transferred in bulk flow systems. The list of these technologies includes baskets, pottery, wheeled vehicles, irrigation systems, horse collars, sails, and rudders. The camel caravans that traveled over trade routes on the Silk Road and around the Indian Ocean from 100 b.c. to 1400 a.d. brought civilizations in Rome, the eastern Mediterranean, East Africa, the Near East, India, southeast Asia, China, and Japan into contact with one another. During this period, the agricultural system that emerged in the Fertile Crescent greatly expanded, and the human diseases that resulted from living in close proximity to domesticated animals traveled over the trade routes to infect distant populations.

After 1492, the transfer of plants, animals, and diseases between Europe and the Americas dramatically altered the global distribution of organisms to a degree that had not been witnessed since the end of the last Ice Age.[22] The introduction of European crops (wheat, rice, sugar, coffee) and animals (cattle, horses, pigs, sheep), and the widespread use of European farming practices massively transformed the natural environment throughout North and South America.[23] The Europeans did not settle in the Americas in great numbers for another three hundred years after the Columbian Exchange began, but during this period they did import large numbers of African slaves. This hideous trade in human beings was deemed necessary because the crops grown in the Americas for sale in Europe required an enormous amount of human labor that was not available in the New World because European diseases had decimated much of the indigenous population.[24]

Two key food staples from the Americas, maize and potatoes, fueled the growth of the European population prior to the Industrial Revolution and massively contributed to its success by vastly increasing the pool of cheap labor.[25] However, the overall rate of increase in the global human population prior to this revolution, from about two hundred and fifty million in 100 a.d. to eight hundred million in 1800 a.d., was modest due to the lethality of crowd diseases and the amount of energy that could be extracted from the earth's biomass by human labor, draft animals, windmills, and waterwheels. But our numbers began to increase dramatically after new technologies made it possible to extract the energy contained in the ancient biomass of the earth as fossil fuels. To better appreciate why this was the case, consider that the energy contained in a barrel of oil is equivalent to that produced by twenty-five thousand hours of human labor or more than a decade of human labor per barrel.[26]

During the nineteenth century, coal-powered steam engines extended the agricultural system that supplied foodstuffs to cities by several orders of magnitude, and the population of cities increased exponentially. After 1860, three developments made it possible to feed the burgeoning population of industrial cities, and each involved prodigious increases in the use of oil. First, a billion acres of new land was incorporated into an increasingly global food production system in the Corn Belt of the United States, in southern Russia, and in the grasslands of Argentina, Australia, and South Africa. Second, the mechanization of agriculture, including the massive use of chemical fertilizers, pesticides, and herbicides, resulted in large increases in agricultural productivity while simultaneously reducing labor costs. And third, new food-preservation technology and a rapidly expanding fossil fuel–based transportation system vastly increased the amount of foodstuffs that could be grown in the Southern Hemisphere and consumed in the cities of Europe and North America.[27]

Until the beginning of the twentieth century, death rates in cities exceeded birthrates due to the lethality of crowd diseases and environmentally induced illnesses, such as heart and lung disease. During this period, cities were death traps, and urban population growth would not have occurred in the absence of a constant influx of healthy peasants from the countryside.[28] In 1800, when the global population was approximately eight hundred million, about twenty-four million people lived in cities, and no city had more than one million inhabitants. In 1900, when the world's population was roughly 1.6 billion, some six hundred million lived in cities, and nine cities had populations greater than one million.[29]

During the twentieth century, the global population increased to more than six billion, about three billion people eventually lived in cities, and twenty-one of these cities now house populations in excess of ten million. This remarkable increase in the numbers of people living in cities could not have occurred in the absence of an increasingly global agricultural production system and a food transportation network that consumed increasingly larger amounts of fossil fuel. Global demand for oil increased so significantly in the twentieth century that at any time after 1970 about five gallons of oil were in transit on large seagoing tankers for every man, woman, and child on the face of the earth. This prodigious flow of oil spurred the phenomenal growth of industrialization worldwide and allowed for the creation and exponential expansion of the petrochemical industry.

By 1999, this industry was producing over a thousand million tons of organic chemicals annually.[30] Degradable materials, such as wood and paper, were replaced with new nondegradable materials, such as plastics. One unfortunate result was that landfills, rivers, streams, and ocean beds become increasingly saturated with the remains of these unwanted but very durable products. The petrochemical industry also manufactured prodigious amounts of chemical fertilizers, herbicides, and pesticides that contributed, along with the heavy use of farm equipment powered by gasoline and diesel engines, to the widespread mechanization and industrialization of agriculture in countries where arable land was plentiful and labor costs were low.

This fossil fueled–based system massively transformed agricultural processes and practices in Europe, North America, Japan, Australia, and New Zealand. Farmers in these countries now specialize in single crops that plant geneticists have bred for responsiveness to chemical fertilizers, resistance to chemical pesticides, and compatibility with mechanized harvesting. The dependence of the new agricultural system on oil was also greatly amplified after patchwork quilt farms were displaced by vast fields dedicated to monoculture. Because monoculture crops are more vulnerable to insects and other pests, a vast increase in the use of chemical pesticides was required to grow them. And because these crops also deplete specific nutrients in the soil much faster, farmers were obliged to vastly increase their use of chemical fertilizers as well.

In the existing global agricultural system, nothing, as the name implies, is local or regional. As David Orr points out, "An Iowa cornfield is a complicated human contrivance resulting from imported oil, supertankers, pipelines, commodity markets, banks and interest rates, federal agencies, futures markets, machinery, spare parts supply systems, and agribusiness companies that sell seeds, fertilizers, herbicides, and pesticides."[31] The foodstuffs produced in this system and the packaged goods and other products made from these materials are now transported over great distances by another system that owes its existence to prodigious supplies of oil—a vast network of barges, ships, cargo vessels, railroads, and trucks. Agricultural crops are typically transported thousands of miles from fields to processing centers, and it is not unusual for the farmers who grow these crops to consume products from these centers that have traveled thousands of miles in the opposite direction to local wholesalers and retailers.

Oil also fuels another transportation system of roughly a billion cars, buses, small trucks, and motorbikes that travel along a network of roads that covers roughly ten percent of the land mass in North America, Europe, and Japan and about two percent of the land mass worldwide.[32] The average American car travels approximately one hundred thousand miles in its lifetime and emits around thirty-five tons of carbon dioxide or monoxide. The more than five hundred million cars in use worldwide generate about twenty-five percent of global greenhouse gas emissions, and the UN Population Fund projects that by 2025 cars in developing countries will be emitting four times as much carbon dioxide as in the industrialized countries today.[33] Also consider that the process of manufacturing a car generates on average about as much air pollution as driving a car for ten years and produces approximately twenty-nine tons of waste.[34]

Since 1960, the rapid emergence of a worldwide telecommunications network allowed systems of production, distribution, and exchange to be linked together into a web of interconnections in a truly global economy. Most of the products that now exist in abundance in retail stores in industrialized countries are assembled at sites around the globe from component parts made in other distant locations. Foodstuffs and other products in grocery stores in these countries travel on average about two thousand miles prior to purchase.[35] And the components of the global economic system (capital, labor, energy resources, raw materials, component parts, finished products, and waste materials) now move through this system with minimal resistance from tariff barriers, transport costs, local markets, and cultural differences.

LESSONS FROM THE NEW STORY

As noted earlier, the most important lesson in this new story of science about how we came to be is that the cause of the environmental crisis and the manner in which it can be resolved are the same. This crisis exists because one species among the millions that have existed on this planet evolved the capacity to acquire and use fully complex language systems. And it can be resolved by using this extraordinary ability to develop and disseminate new narratives about political and economic reality that can serve as the basis for implementing scientifically viable solutions for environmental problems.

This story also challenges the validity of two assumptions in the so-called standard social science model still widely used by historians, literary critics, anthropologists, sociologists, and psychologists. The first assumption is that the human brain is an infinitely pliable blank slate that serves as the basis for recording narratives in particular cultural contexts. The second is that the cultural narratives assimilated or learned in diverse cultures result in constructions of reality that are utterly different from one another. If these assumptions were valid, there would be no prospect for achieving the totally unprecedented level of good will and cooperation between peoples and governments required to resolve the environmental crisis. Fortunately, we now know they are not valid.

Research in neuroscience and the behavioral sciences has shown that the brains of all normal human infants are remarkably similar and massively condition language acquisition, cognitive abilities, modes of perception, and a wide range of behaviors. Recent studies in cognitive science have demonstrated that mental structures in the human brain result in cultural universals that are hidden beneath the maze of cultural diversity. And the list of these cultural universals ranges from color and number codes to facial expressions to courting behavior to conceptions of territoriality and the structures of legal systems.[36]

Another lesson from the new story of science that could be hugely important in the effort to resolve the environmental crisis involves research in neuroscience made possible by computer-based, brain-imaging systems. This research has revealed that the mirror neuron system not only allows us in our interactions with others to share emotions and grasp intentions on a prereflective level, or in the absence of higher level cognitive functions associated with language use. It also makes it possible to experience the other as oneself regardless of language barriers, religious beliefs, and political ideologies. Equally important, research on the neuronal processes that result in spontaneous moral behavior has shown that the human capacity to treat the other as one would wish to be treated is innate. And this conclusion has been reinforced by the research in behavioral and social sciences which strongly suggests that the emotions and moral concepts associated with this behavior are universal.

This does not mean that we are obliged to conclude, as the poet E. E. Cummings put it, that "since feeling is first, who cares about the syntax of things?" It does mean, though, that the syntax of things in the discourse we now use in the effort to resolve environmental problems is badly in need of revision. This discourse is replete with quantitative reasoning, risk assessments, cost-benefit analyses, and technological fixes that tell us that a problem exists but do not provide compelling reasons for why it should be resolved. And it also precludes the prospect of engaging in the implicit, automatic, and prereflective human interactions that result in a sense of shared purpose and common understanding.

The next chapter considers what the new story of physics has to say about the relationship between the trillions of atoms intricately assembled in highly specialized ways in the part we call self and the whole that is the cosmos. For those readers who may be having a mild anxiety attack when faced with the prospect of studying physics, fear not. All that will be required to understand the material on the brave new world of modern physics is a willingness to exercise your imagination and suspend belief in common sense assumptions about physical reality that are not commensurate with the actual character of this reality.

CHAPTER 3

The New Story in Physics: Mind, Matter, and the Nonlocal Universe

The paradox is only a conflict between reality and what you think reality ought to be.

>Richard Feynman

Unless all ages and races of men have been deluded by some mass hypnotist (who?), there seems to be such a thing as beauty, a grace wholly gratuitous.

>Annie Dillard

The capacity to acquire and use fully complex language systems made the members of our species fully conscious and self-aware beings in the vast cosmos. But this enormous privilege came with a price. After our ancestors began to live storied lives in a linguistically based symbolic universe, the world that previous generations experienced as an integrated and undivided whole split into two worlds—an inner world where the self that is aware of its own awareness exists and an outer world in which this self seeks to gratify its needs and establish a meaningful sense of connection with other selves. And this explains why the most fundamental impulse in the storied lives of fully modern humans has always been to close the gap between these inner and outer worlds by integrating all seemingly discordant parts of a symbolic universe into a meaningful and coherent whole.

The narrative that has consistently served this function is religion. But during the first scientific revolution of the seventeenth century, another narrative emerged called Newtonian or classical physics that also promised to bridge the gap between self and world by integrating all of the seemingly discordant parts of the physical universe into a coherent and meaningful whole. In this physics, one universal force, gravity, governs the motion, interaction, and blending of indestructible atoms or mass points. And since the laws of gravity were completely deterministic, it was assumed that all events in the cosmos are predetermined by the forces associated with these laws and that the future of any physical system could be predicted with absolute certainty if initial conditions are known.

In the worldview of classical physics, human beings were cogs in a giant machine and linked to other parts of this machine in only the most mundane material terms. The knowing self was separate, discrete, and isolated from the physical world, and all the creativity of the cosmos was exhausted in the first instant of creation. As physicist Henry Stapp points out, "Classical physics not only fails to demand the mental, it fails to even provide a rational place for the mental. And if the mental is introduced ad hoc, then it must remain totally ineffectual, in absolute contradiction to our deepest experience."[1]

During the second scientific revolution of the twentieth century, a new story in physics emerged that challenged and effectively undermined all of the assumptions about physical reality in classical physics. In this story, the cosmos is a sea of energy in which more complex systems spontaneously emerge that display novel properties that cannot be reduced to or explained in terms of their constituent parts. Matter cannot be dissected from this omnipresent sea of energy, and it is not possible to observe matter from the "outside."

Even more remarkable, the new story of physics revealed that the universe is a single significant whole in which all processes are interconnected, interrelated, and interdependent at all scales and times. As physicist Werner Heisenberg put it, the cosmos "appears as a complicated tissue of events, in which connections of different kinds alternate or overlay or combine and thereby determine the texture of the whole."[2] The new story of physics does not suggest that our existence was planned or preordained by God or any other spiritual agency. But it does provide a place for the knowing mind and a basis for believing that human life and consciousness are embedded in and emergent from a self-organizing and self-perpetuating cosmos at a very high degree of complexity.

This new story begins in 1905, when Albert Einstein, then an obscure patent clerk working in Geneva, Switzerland, published the special theory of relativity. In classical physics, space and time existed in separate and distinct dimensions of physical reality, and it was possible to view a frame of reference where an observation is made as absolutely at rest. This view of space and time also made it possible to assume that material objects could achieve a velocity equal to the speed of light and that an observer traveling at this velocity would perceive the speed of light as zero.

Einstein's first insight was that there was no basis for determining absolute motion, or motion that proceeds in a fixed direction at a constant speed, in the absence of a comparison with the motion of other objects. The second was that there is no frame of reference absolutely at rest and that the only constant for observers in all frames of reference is the speed of light. Based on these two postulates, the relativity of motion in different frames of reference and the constancy of the speed of light, the entire logical structure of the special theory of relativity followed. In this theory, Einstein mathematically deduced the laws that related space and time measurements made by one observer to those made by another observer moving uniformly relative to the first.

Because the speed of light in the special theory of relativity is constant for all observers, it does not vary regardless of the relative velocity of the reference frame in which the observation is made. If, for example, one could chase a photon, or particle of light, at increasingly greater velocities in a spaceship, it would still be moving away at its own constant speed. The reason why these relativistic effects are not obvious, said Einstein, is that the speed of light is so fast in comparison to ordinary speeds that this creates the illusion that we perceive events in the instant they occur.

In the space-time description Einstein used to account for observational differences between frames of reference moving at different velocities, time is a coordinate along with three space coordinates in a four-dimensional space-time continuum. And transformations between different frames of reference express each coordinate of one frame as a combination of the coordinates of the other frame. A space coordinate in one frame usually appears as a combination, or mixture, of space and time coordinates in another frame, and measuring instruments change from one frame of reference to another. This means that simultaneous events in these frames of reference would appear to occur at different times and that two clocks moving in frames of reference with different relative velocities would not register the same time.

For example, if you were standing on a platform watching a train one hundred yards long moving through the station at sixty percent of the speed of light, it would appear to be only eighty yards long. If you could hear the conversations of the people on the train and watch their actions, their voices would sound like a record played at too slow a speed, and they would appear to moving in slow motion. Because the mixture of space-time coordinates in the reference frame of the train traveling at this enormous velocity is very different from that on the station platform, the clocks on the train would seem to be running at about four-fifths of their normal speed. If your identical twin were traveling on the train, his or her biological clock would be running more slowly than your own, and he or she would be younger that you are when the train stopped in the station. But the twin on the train would not be aware of any of these relativistic effects, and life would seem perfectly normal.

The special theory of relativity dealt only with constant motion of the frames of reference, and Einstein extended the framework of this theory in the general theory of relativity (1915) to account for the more general case of accelerated frames of reference. Einstein's seminal insight here was that it is impossible to distinguish between the effects of gravity and nonuniform motion. If you did not know, for example, that you were on an accelerating spaceship and dropped a cup of coffee, it would not be possible to determine if the mess on the floor was due to the effects of gravity or the accelerated motion. This inability to distinguish between a nonuniform motion, like an acceleration, and gravity is known as the principle of equivalence. Based on this principle, Einstein in the general theory posited the laws relating space and time measurements carried out by two observers moving uniformly in accelerated frames of reference. In this theory, the force field of gravity causes space-time to become warped or curved, and the motion of objects is along geodesics in curved space (see figure 3.1).

What is most important for the purposes of this discussion is that the general theory of relativity disclosed a startling new relationship between the part of physical reality we call self and the whole that is the cosmos. In the general theory, wrote physicist Max Planck, "Each individual particle of the system in a certain sense, at any one time, exists simultaneously in every part of the space occupied by the system."[3] And this system, as Planck points out, is the entire cosmos. As

Figure 3.1
General Theory: Space-Time Curvature

Einstein put it, "Physical reality must be described in terms of continuous functions in space. The material point, therefore, can hardly be conceived any more as the basic concept of the theory."[4]

Since the human body in Einstein's view is a collection of material particles, he concluded that any sense we might have that our physical self is separate from the world is an illusion fostered by a lack of understanding of the actual character of physical reality. "A human being," said Einstein, "is a part of the whole, called by us the 'Universe,' a part limited in time and space. He experiences himself, his thoughts and feelings as something separate from the rest—a kind of optical illusion of his consciousness. This delusion is a kind of prison for us, restricting us to our personal desires and to affection for a few persons nearest to us. Our task must be to free ourselves from the prison by widening our circle of compassion to embrace all living creatures and the whole of nature in its beauty. Nobody is able to achieve this completely, but the striving for such achievement is in itself a part of the liberation and a foundation for inner security."[5]

THE STRANGE NEW WORLD OF QUANTUM PHYSICS

This radically new understanding of the relationship between self and world was massively reinforced during the 1920s in relativistic quantum field theory. Physical reality in this theory is emergent from and embedded in interactions between matter-like entities called quanta and immaterial force fields, and quanta manifest as *either* waves or particles. In classical physics, it was possible to assume that all properties of a physical system, including those of atoms and molecules, were exactly definable and determinable and that the future of any physical system could be predicted with absolute certainty. As French mathematician Pierre-Simon de Laplace famously put, "An intellect which at a given instant knew all forces acting in nature and the position of all things of which the universe consists, would be able to comprehend the motions of the largest bodies of the world and those of the lightest atoms in one single formula, provided that this intellect were sufficiently powerful to subject all data to analysis; both past and future would be present to his eyes."[6]

But in the new story of physics, scientific knowledge is always proximate, nothing can be known with absolute certainty, and the language of mathematical theories cannot fully disclose and describe physical reality. One reason why this is the case is that wave-particle dualism makes it impossible to determine definite values of physical systems on the quantum mechanical level in the absence of measurement. Another is that the future state of these systems can be predicted only within a range of probabilities and a proper understanding of observed values must take into account the presence of the observer and the measuring instruments.

Picture the particle as a point-like entity, like the period at the end of this sentence, and the wave as a multidimensional entity spread out in all directions at once, like a water wave created when a stone is thrown into a pond (see figure 3.2).

The wave aspect of quanta results in the formation of interference patterns which are analogous to the interference patterns in water waves that should be familiar to anyone who has swum in the ocean. These patterns result when two waves produce peaks in places where they combine and troughs where they cancel each other out (see figure 3.3).

The wave function in quantum physics is described by the Schrodinger wave equations. These equations are completely deterministic and allow us, in theory at least, to predict the future of a quantum system with certainty. But when a quantum system is observed, the wave function does not allow us to predict where the particle will appear at a specific location in space. It only allows is to predict the probability of finding the particle within a range of

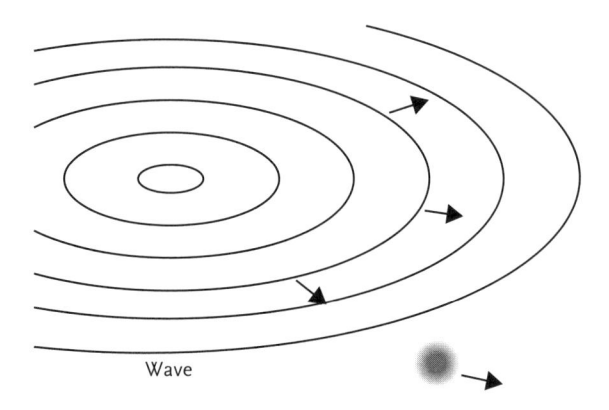

Figure 3.2
Wave and Particle Picture

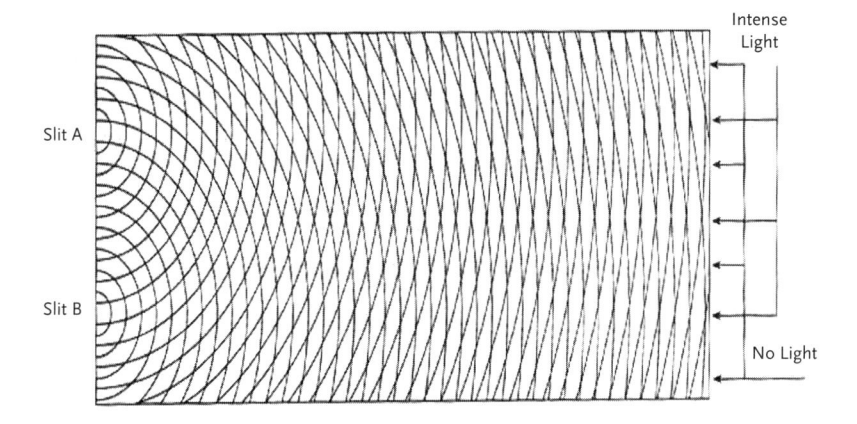

Figure 3.3
Wave-Interference Patterns

probabilities associated with all possible states of the wave function (see figure 3.4).

In order to get a better sense of what the new story of physics has revealed about the actual character of physical reality, imagine that the universe is a 3-D movie. The projectors in this movie are the four immaterial force fields—strong, electromagnetic, weak, and gravitational. Potential vibrations at any point in these projector fields are capable of generating messenger quanta—the graviton for gravity, the photon for electromagnetism, the intermediate bosons for the weak force, and the colored gluons for the strong force. And the exchange of messenger quanta within and between these fields results in the matter quanta that constitute physical reality.

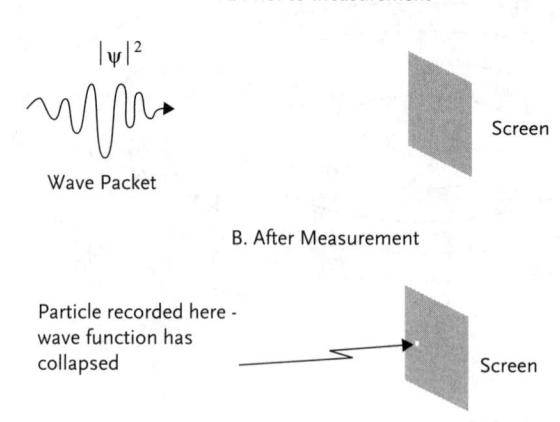

Figure 3.4
Collapse of Wave Function

In our cosmic 3-D movie, quanta are not discrete and separate entities acted upon by external forces like atoms or masses in classical physics. They emerge from a web of interaction with other particles, and the only action that we can observe, detect, or measure results from these interactions. As Henry Stapp puts it, "Each atom turns out to be nothing but the potentialities in the behavior pattern of others. What we find, therefore, are not elementary space-time realities, but rather a web of relationships in which no part can stand alone; every part derives its meaning and existence only from its place within the whole."[7]

The act of making observations or measurements can be likened to putting on the glasses required to see the action in the 3-D movie. And the action that we might presume to be there in the absence of measurement, or before putting on the glasses, is not the same as that which we actually observe. This is the case because quanta have two logically antithetical but complementary aspects, wave and particle, and it is not possible to perceive or measure both simultaneously.

One of the easiest ways to understand the special character of wave-particle dualism is to examine the results of the famous two-slit experiment. As physicist Richard Feynman put it, "Any other situation in quantum mechanics, it turns out, can always be explained by saying, 'You remember the case with the experiment with two holes? It's the same thing.'"[8] In our idealized two-slit experiment we have a source of quanta, an electron gun like

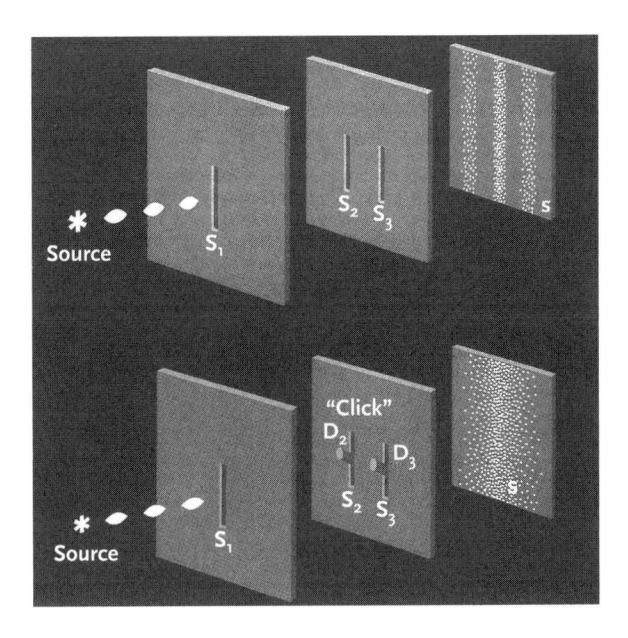

Figure 3.5
Two-Slit Experiment

that in conventional television sets, and a screen with two openings that are small enough to be comparable with the wavelength of an electron. Our detector is a second screen that flashes, like those in conventional television sets, when an electron impacts it (see figure 3.5).

When both slits S_2 and S_3 are open, each slit becomes a source of waves. The waves spread out spherically, come together, and produce interference patterns that appear as bands of light and dark on the detector. Dark stripes reveal where the waves have canceled each other out, light stripes where they have reinforced one another. If one of the openings is closed, a bright spot appears on the detector in line with the other slit, indicating that the electrons have traveled as particles through this slit like bullets.

This experiment has been conducted with a single electron and its associated wave packet arriving at a detector one at a time. If we view an electron as a particle, or a point like something with a specific location in space and time, it seems reasonable to assume that it would move through one slit or the other when both are open. But if we conduct this experiment many times with both slits open, we see a buildup of the interference patterns, which indicates that the single particle has behaved like a wave (see figure 3.6).

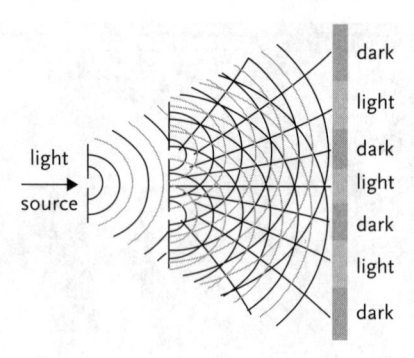

Figure 3.6
Results with Two Slits Open

Now suppose we refine our experiment in an attempt to determine which of the two slits an electron passes through by putting a detector (D_2 and D_3) at each slit (S_2 and S_3) (figure 3.5). After many electrons pass through the slits, we discover bright spots in direct line with each opening where the firing of a detector indicates that the electron in its particle aspect has passed through. Since no interference patterns associated with the wave aspect are observed, the electrons are behaving like particles. But we cannot predict which detector will fire or which slit the electron as particle will pass through. All we can know is that there is a fifty percent probability that it will travel through one slit or the other each time we run the experiment.

Let us now conduct a two-slit experiment designed to disclose both the wave and particle aspects of a single photon or particle of light by making extremely rapid changes in our experimental apparatus. In this experiment, photographic plates are placed behind the two slits, which function like venetian blinds. When the blinds are closed, interference patterns associated with the wave aspect of the photon will appear on the detection screen. When they are opened, a bright spot will appear on the detection screen in line with one of the slits, indicating that the photon is behaving like a particle. And each time we run our experiment, there is a fifty percent probability that the single photon will travel though one slit or the other.

Now suppose we open or close the venetian blind-like plates "after" a single photon has traveled through the slits. This arrangement was originally proposed in 1978 by physicist John A. Wheeler in what is known as the delayed-choice thought experiment. According to the predictions of this thought experiment, interference patterns associated with the wave aspect of the photon will appear on the

Rebirth of the Sacred

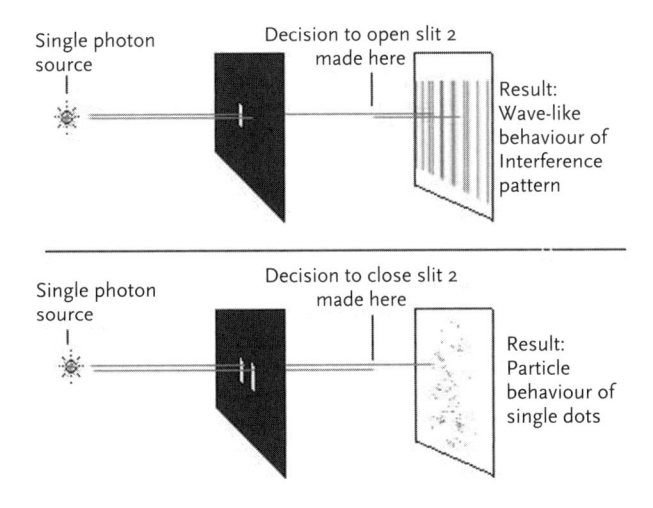

Figure 3.7
Delayed-Choice Thought Experiment

detection screen when the blinds are closed "after" the photon in its particle aspect has passed through both slits. This experiment also predicts that if both blinds are opened "after" a photon in its wave aspect has passed through the slits when there were closed, a detector at one of the slits will fire and reveal the presence of the particle aspect. The quantum strangeness here is not merely that we cannot predict which detector will fire. It is also that we are determining the state of the photon with an act of observation "after" it has passed through the slits. As Wheeler puts it "we decide, after the photon has passed through the screen, whether it shall have passed through the screen"[9] (see figure 3.7)

Wheeler's prediction that a single photon will follow two paths, or one path, according to a choice made after the photon has followed one or both paths has been repeatedly confirmed in actual delayed-choice thought experiments.[10] The results of these experiments not only demonstrate that the observer and the observed system cannot be separate and distinct in space. They also reveal that this distinction does not exist in time. It is as if we caused something to happen after it already occurred. These experiments also revealed another strange feature of the quantum world—the past is inexorably connected with the present, and even the phenomenon of time is tied to specific experimental choices.

The obvious question here is what could these esoteric experiments with electrons and photons possibly say about the world in which we live? The answer is that they disclose general properties of quanta and, therefore, fundamental aspects of everything that exists in physical reality. Since quantum mechanical events cannot be directly perceived by the human senses, experience does not teach us that physical reality emerges from the interactions between fields and quanta. But this would be quite obvious if you were reduced in size about twenty orders of magnitude and could observe the quantum mechanical events that result in the emergence of matter quanta, atoms, and molecules. The world as you would then perceive it would be a web of interactions between fields and quanta in which nothing is separate and distinct and everything is seamlessly connected with everything else. If one could see the real world on this level, the universe would be perceived, as Brian Swimme and Thomas Berry put it, "not as a thing, but a mode of being everything."[11]

A NEW FACT OF NATURE

In the strange new world of quantum physics, we have consistently uncovered aspects of physical reality that are at odds with our everyday sense of this reality. But no previous discovery has posed more challenges to our usual understanding of the "way things are" than an amazing new fact of nature known as nonlocality. Perhaps the best way to appreciate why this is the case is to imagine that you are one of two observers in a scientific experiment. In this experiment, two photons originate from a single source and travel in equal and opposite directions halfway across the known universe to points where each will be measured or observed. Now imagine that before the photons are released, each of two observers is magically transported to one of the points of observation. The task of the observers is to measure and record a certain property of each photon that has traveled over this enormous distance to their point of observation.

In spite of the fact that the photons are traveling at the speed of light, each observer would have to wait patiently for billions of years for one of the photons to arrive at his or her observation point. Suppose, however, that the observers are willing to endure this wait because they hope to test the predictions of a mathematical theorem. This theorem not only allows for the prospect that there could be a correlation between the observed properties of the two photons. It also indicates that this correlation could occur instantly, or in no time, in spite of the fact that the distance between the observers and their measuring instruments is billions of light years.

Now imagine that after the observations are made, the observers are magically transported back to the point where the photons were released and the observations recorded by each are compared. Strangely enough, the results would indicate that the observed properties of the two photons did, in fact, correlate with one another over this vast distance instantly or in no time. This imaginary experiment distorts some of the more refined aspects of the actual experiments in which photons released from a single source are measured or correlated over what physicists term space-like separated regions. But if we assume that the imaginary experiment was conducted many times, the results would be the same of those in the actual experiments.

The intent in these experiments was to test some predictions made in a mathematical theorem published in 1964 by physicist John Bell. Like Einstein, Bell was discomforted by the threat quantum physics posed to the assumption in classical physics that a physical theory could predict the future of a physical system with absolute certainty. And he hoped that the results of the experiments testing his theorem would obviate this threat. But Bell was also aware that these results would provide a basis for answering some very large questions about physical reality and the ability of physical theory to disclose and describe this reality.

Would the results reveal that quantum physics is a self-consistent theory whose predictions would hold in this new class of experiments? Or would they reveal that quantum theory is incomplete and that its apparent challenges to the assumption that physical theories can predict the future of physical systems with absolute certainty are illusory? Answers to these questions derived from the experiments testing predictions made in Bell's theorem would also determine which of two fundamentally different assumptions about the character of physical reality is correct. Is physical reality, as classical physics assumed, local? Or is physical reality, as quantum theory predicts, nonlocal? While the question may seem esoteric, and the terms innocuous, the issues at stake and the implications involved were enormous.

Bell was convinced that the totality of all of our previous knowledge of physical reality, not to mention the laws of physics, would favor the assumption of locality. This assumption states that a measurement at one point in space cannot influence what occurs at another point in space if the distance between the points is large enough that a signal cannot travel between them at light speed in the time allowed for measurement. But quantum theory allows for what Einstein disparagingly termed "spooky actions at a distance."

When two particles originate under conditions where their wave aspects are entangled, quantum theory predicts that a measurement of a property of one particle will correlate with that of the other particle even if the distance between them is billions of light years. The theory also indicates that even though no signal can travel faster than light, the correlations will occur instantaneously or in "no time." If this prediction held in experiments testing Bell's theorem, we would be forced to conclude that physical reality is nonlocal.

After Bell published his theorem in 1964, a series of increasingly refined tests of the predictions made in this theorem culminated in experiments by Alain Aspect and his team at the University of Paris-South. When the results of the Aspect experiments were published in 1982, the answers to Bell's questions were quite clear—quantum physics is a self-consistent theory, and nonlocality is a fact of nature.[12] In 1997, these same answers were provided by the results of twin-photon experiments carried out by Nicolus Gisin and his team at the University of Geneva.[13]

While the distance between detectors in the Aspect experiments was 13 meters, the distance between detectors in the Gisin experiments was extended to 11 kilometers or roughly seven miles. This distance is so vast in the realm of quanta that if the strength of the correlations held at 11 kilometers, physicists were convinced they would also hold in an experiment where the distance between the detectors was halfway to the edge of the entire universe. But if the strength of the correlations significantly weakened or diminished, this would indicate that physical reality is local and that nonlocality does not apply to the entire universe. This did not prove to be the case. The results of the Gisin experiments revealed that the correlations between detectors located in these space-like separated regions did not weaken as the distance increased.[14]

What is most important to realize here is that the new fact of nature revealed in these experiments, as physicist Bernard d'Espagnat was among the first to point out, is a "general" property of nature.[15] All particles in the history of the cosmos have interacted with other particles in the way that particles interact in the Aspect and Gisen experiments. Virtually everything in our immediate physical environment is made up of quanta that have been interacting with other quanta in this manner from the big bang to the present. Also consider that quantum entanglement grows exponentially with the number of particles that interact under conditions where this can occur and that there is no theoretical limit on the number of these entangled particles.[16] This means that on a very basic level the universe is a vast web of particles that remain in contact with one another over any distance in no time and that one point in space-time is in some sense every point in space-time.[17]

LESSONS FROM THE NEW STORY

In the new story of physics, the most fundamental scientific truth is that physical reality is a single significant whole and that everything within the vast cosmos is emergent from and embedded in this whole. As Erwin Schrodinger put it, "Hence this life of yours which you are living is not merely a piece of the entire existence, but is, in a certain sense, the whole; only this whole is not so constituted that it can be surveyed in one single glance."[18] For the purposes of this discussion, this is extremely significant. And the reason why this is the case is that the most profound religious truth in the great religious traditions of the world is that spiritual reality is a single significant whole and that human life and consciousness are emergent from and embedded in this whole.

Another equally profound religious truth in these traditions is that the single significant whole in spiritual realty cannot be reduced to or fully understood in ordinary language. And this truth is analogous to and entirely compatible with the fundamental scientific truth in the new story of physics that the whole of physical realty cannot be reduced to or fully understood in mathematical language. This should have been obvious when Kurt Gödel published his incompleteness theorem in 1930. This extremely important but often ignored theorem shows that mathematics, the language of physical theory, cannot reach closure because no algorithm, or calculation procedure, that uses mathematical proofs can prove its own validity. This means that any mathematical description of physical reality that claims to exhaustively describe any aspect of this reality cannot prove itself. As physicist Freeman Dyson puts it, "Gödel proved that in mathematics the whole is always greater than the sum of the parts."[19]

According to mathematician Rudy Rucker, "Mathematics is open-ended. There can never be a final best system of mathematics. Every axiom-system for mathematics will eventually run into certain simple problems that it cannot solve at all."[20] If a mathematical system cannot prove itself, it follows that mathematical systems in physical theories will never be able to fully disclose or describe the actual dynamics of physical reality. Even if we did manage to develop what physicists call a Theory of Everything, this theory could not in principle be the final or complete description.

It was quantum physicists and not philosophers or theologians who first realized that there is a new basis for dialogue between the truths of science and religion. Schrodinger expressed his hope that this dialogue would eliminate what he viewed as a terrible deficiency in the scientific worldview: "The scientific picture of the real world around me is very deficient. It gives me a lot of factual information, puts all our experience in a magnificently consistent order, but it is ghastly silent about all and sundry that is really dear to our hearts, that really matters to us."[21]

Physicist Wolfgang Pauli, who also made substantive contributions to the development of quantum theory, said he was hopeful that the new dialogue between the truths of science and religion would resolve another big problem: "Contrary to the strict division of the activity of the human spirit into separate departments—a division prevailing since the nineteenth century—I consider the ambition of overcoming opposites, including also a synthesis embracing both rational understanding and the mystical experience of unity, to be the mythos, spoken and unspoken, of our present day and age."[22] The "strict division" of science and religion into "separate departments" has been one of the principle impediments to the active involvement of people of faith in efforts to resolve the environmental crisis. But as Pauli realized decades ago, the new story of physics resolves this problem because it provides a basis for achieving a synthesis between a rational scientific understanding of the universe and the "mystical experience of unity."

The next chapter will demonstrate that the understanding of the relationship between parts (organisms) and whole (biosphere) in the new story of biology is remarkable similar to that in the new story of physics. This chapter will also provide a coherent basis for understanding why the political and economic narratives that now serve as the basis for coordinating global human activities are predicated on unscientific assumptions about the relationship between human and environmental systems. One large problem here is that the unscientific assumptions in these narratives effectively preclude the prospect of implementing scientifically viable solutions for problems in the global environment in binding international agreements. The other is that the business as usual approach to resolving the environmental crisis embraced by virtually all political leaders and economic planners is predicated on these assumptions.

CHAPTER 4

The New Story in Biology: Parts and Wholes in the Web of Life

Overpopulation, the destruction of the environment, and the malaise of the inner cities cannot be resolved by technological advances, nor by literature or history, but ultimately only by measures based on an understanding of the biological processes involved.

Ernst Mayr

In July of 1969 the Apollo 11 spacecraft emerged from the dark side of the moon and the on-board camera panned through the vast emptiness of outer space. Against the backdrop of interstellar night hung the great ball of earth, with the intense blue of its oceans and the delicate ochres of its landmasses shimmering beneath the vibrant and translucent layer of its atmosphere. In the shock of this visual moment, distances between us contracted; boundaries and borders ceased to exist. But the impression that sent the adrenaline flowing through my veins was that the teeming billions of organisms writhing about under the protective layer of the atmosphere were not separate—they were interdependent, fluid, and interactive aspects of the one organic dance of the planet's life.

The preceding paragraph, an entry form my diary written a few days after images of the whole earth first appeared on television, cannot be classed as scientific analysis. But it is entirely consistent with what the new story of science has revealed about the relationship between human and environmental systems in biological reality. The large problem here is that the political and economic narratives that now serve as the basis for coordinating global human activities are premised on scientifically outmoded assumptions about this relationship in the old story of classical physics. And this problem is further complicated by the fact that the view of this relationship that is still widely viewed as scientific in Darwin's theory of evolution is also premised on these scientifically outmoded assumptions.

Darwin went public with his theory for the first time in a paper presented to the Linnean Society in 1848. This paper begins with the following sentence: "All nature is at war, one organism with another, or with external nature."[1] In *The Origins of Species*, Darwin is more specific about the character of this war: "There must be in every case a struggle for existence, either one individual with another of the same species, or with the individuals of distinct species, or with the physical conditions of life."[2]

Based on the assumption that the study of variation in domestic animals and plants "afforded the best and safest clue" to understanding evolution, Darwin concluded that nature could by cross-breeding and the selection of traits produce new species. The explanation of how this happens took the form of the following argument: (1) more individuals in each species are produced than can survive; (2) the struggle for existence results from a shortage of resources; (3) organisms that live long enough to reproduce have slight variations that provide a survival advantage; and (4) the variations that provide this advantage are passed on to subsequent generations and accumulate in interbreeding groups that over time can become new species. In an analogy with the animal breeder's artificial selection of traits, Darwin named this process natural selection.

Darwin's struggle for existence occurs *between* an individual organism and other individual organisms in the "same species," *between* an individual organism of one species and that of a "different species," or *between* an individual organism and the "physical conditions of life." The parts in this theory resemble classical atoms, and the force that drives the interactions of the parts, the "struggle for life," resembles Newton's force of universal gravity. Darwin's claim that mutations within organism are random may have challenged the assumption in classical physics that the future of physical systems can be predicted with ultimate certainty. But his understanding of the force that drives the process of evolution and how this force governs the interactions between organisms was clearly Newtonian.

Darwin did not say anything about the mechanisms within organisms that result in "slight variations" because the concept of genes was not known to him. A contemporary of Darwin, the monk Gregor Mendel, introduced this concept in a paper published in an obscure Austrian journal in 1865. But it was not until about 1910 that scientists began to realize that genes could be the basic unit of evolution and that a better understanding of gene transmission could put the theory of evolution on firmer scientific foundations.

The new story of biology begins in the 1940s, when Theodosius Dobzhansky, Ernst Mayr, and others synthesized Darwin's idea that organisms and populations gradually change with Mendel's concept of genetic inheritance in what became known as the Modern Synthesis. The aspect of this synthesis that most directly challenged the mechanistic assumptions about the relationship between parts (organisms) and whole (biosphere) in Darwin's theory of evolution is the concept of emergence. The term *emergence* applies to situations where new wholes spontaneously appear that have properties or display behavior that cannot be reduced to or understood in terms of their constituent parts. As Mayr put it, living systems "almost always have the peculiarity that the characteristics of the whole cannot (not even in theory) be deduced from the most complete knowledge of components, taken separately or in other partial combinations. This appearance of new characteristics in wholes has been designated emergence."[3]

In the new story of biology, the evolution of more complex organisms and biological processes occurs when an assemblage of parts in successive levels of organization results in new wholes that display novel properties or behavior. "Each higher level subject," notes biologists P. B. and J. S. Medawar, "contains ideas and conceptions peculiar to itself. These are the 'emergent' properties."[4] From this perspective, organisms are not mixtures or compounds of inorganic parts, but new wholes with emergent properties that are embedded in and intimately related to other organisms with their own emergent properties.

The systems that display these properties are nonlinear, and they exist in a state or condition known as far-from-equilibrium. A physical system is in equilibrium when its energy is distributed in the most statistically probable way and change tends not to occur because forces, influences, and reactions cancel each other out. In systems far-from-equilibrium, energy is not evenly distributed and dramatic changes tend to occur because the resulting disequilibrium greatly amplifies forces, influences, and reactions.

Over the past three decades, scientists have used nonlinear systems theory to study emergent properties in far-from-equilibrium systems with the use of increasingly more advanced computer systems. This project began in the 1960s, when Nobel laureate Ilya Prigogine used the computational tools of nonlinear systems theory and the mathematics of complexity to develop a nonlinear thermodynamics for far-from-equilibrium systems. Prigogine's theory revealed that nonlinear systems in far-from-equilibrium interact with other systems via self-reinforcing positive feedback loops. The theory also revealed that these feedback loops can result in new patterns or behaviors that are not imposed by any external agency in a process Prigogine termed "spontaneous self-organization."

When a nonlinear system arrives at the point where the self-reinforcing positive feedback loops can trigger spontaneous self-organization, there are a number of paths or branches that can be followed. And the path or branch that is followed is unique for each system and cannot be predicted with any degree of certainty. This means that there is an irreducibly random element in biological processes that makes it impossible to predict future conditions with absolute certainty. For example, it will never be possible to make precise predictions about weather conditions in the nonlinear system of the atmosphere beyond a limited time period. The current estimate is ten days or less.

Darwin was a brilliant scientist, and his understanding of the dynamics of evolution laid the conceptual foundations for the revolution in the biological sciences that began with the Modern Synthesis. But there is no basis in Darwin's theory for even entertaining the idea that organisms can display spontaneous self-organization that results in new patterns and behaviors. In his theory, parts (organisms) are separate and discrete entities; the force that drives the evolutionary process, the "struggle for life," acts outside or between the parts; and the whole (biosphere) is the sum of the external interactions between the parts. There is nothing in this mechanistic paradigm that could even begin to explain how new patterns and behaviors could spontaneously emerge in organisms in the absence of an agency outside these organisms that imposes these structures or patterns. And there is no basis in this paradigm for viewing organisms as existing in a dynamic web of embedded interactions with their environments that makes it impossible to make categorical distinctions between processes that operate inside and outside these organisms.

PARTS AND WHOLES IN THE SYSTEM OF LIFE

The new story of science has revealed that the process of emergence has resulted in higher levels of organization and more complex physical processes at all scales and times in the history of the cosmos. On the most basic level of physical reality, the interactions between quanta within and between fields result in fundamental particles with emergent properties that do not exist in the quanta. The fundamental particles interact with other fundamental particles to produce the roughly one hundred naturally occurring elements that have emergent properties not found in the particles. The interactions between the elements result in compounds and minerals that have emergent properties not present in the elements. The interactions between the compounds and minerals resulted in the emergence of the ancestor of DNA, which displayed properties that did not exist in the compounds and minerals. And the process of emergence during the entire history of the evolution of life on earth resulted in increasingly more complex life forms and biological processes.

This process began with one self-replicating molecule, the ancestor of DNA, and all of the organisms that have existed on this planet are its direct descendents.[5] During the first two billion years prokaryotes, or organisms composed of cells with no nucleus, were the only living inhabitants of planet earth. But over the course of these two billion years, interactions between these simple organisms resulted in the emergence of the complex processes of fermentation, photosynthesis, and oxygen breathing.

The reason why interactions between these simple organisms could result in such complex processes is that the absence of a nucleus with a surrounding membrane in the prokaryotes allowed bits of genetic material to be routinely and rapidly transferred between them. This allowed individual prokaryotes or bacteria to use accessory genes, often from very different strains, which could perform functions that were not performed by their own DNA. What this picture suggests, according to Lynn Margulis and Dorion Sagan, is that "all the world's bacteria have access to a single gene pool and hence to the adaptive mechanisms of the entire bacterial kingdom."[6] This explains why the biosphere could adapt to changes in the global environment in a few years. As Margulis and Sagan put it, "By constantly and rapidly adapting to environmental conditions, the organisms of the microcosm support the entire biota, their global exchange network ultimately affecting every living plant and animal."[7]

Symbiotic alliances between organisms that became permanent is another aspect of our new understanding of biological reality that is not in accord with the assumption in Darwin's theory that natural selection acts outside or between organisms. For example, there are several organelles, or subunits that have special functions and double membranes, in the cells of higher organisms that were originally separate organisms. The mitochondria found in the cytoplasm of modern animal cells allow these cells to utilize oxygen and to exist in an oxygen-rich environment. But the mitochondria resemble separate organisms because they have their own genes composed of DNA, reproduce by simple division, and do so at times different from the rest of the cell.

The explanation for this extraordinary alliance between mitochondria and the rest of the cell is that oxygen-requiring prokaryotes in primeval seas combined with other prokaryotes. These ancestors of modern mitochondria provided waste disposal and oxygen-derived energy in exchange for food and shelter, and the previously separate organisms evolved together into more complex forms of oxygen-requiring life. Similarly, the ancestors of the chloroplasts inside the cells of all green plants were originally separate organisms that evolved the capacity to convert carbon dioxide and water into oxygen and sugar in the process of photosynthesis. Like the ancestors of the mitochondria, these prokaryotes combined with other prokaryotes, and a new life form emerged as a result of this symbiotic relationship.[8]

Even more remarkable, a great deal of evidence suggests that self-regulating and self-perpetuating dynamics in the whole (biosphere) emerged from the interactions of the parts (organisms). For example, the fossil record indicates that the temperature of the earth's surface and the composition of the air have been continuously regulated by the interactions between organisms. Although the complex network of feedback loops that maintains conditions suitable for life is not well understood, there are good reasons to believe that the entire biota was responsible.

We also know that the concentrations of atmospheric oxygen stabilized at about twenty-one percent millions of years ago and have stayed at this level ever since. If the concentrations had fallen a few percent below this level, oxygen-breathing organisms would have died from affixiation. And if the concentrations of this volatile gas had risen a few percent above twenty-one percent, this would have caused living organisms to spontaneously combust. The only reasonable explanation why this did not happen is that feedback loops emerged in the interactions between organisms that regulated the concentrations of oxygen and other atmospheric gases.[9]

Also consider that the total luminosity of the sun, or the total quantity of energy released as sunlight per year, has increased during the past four billion years by as much as fifty percent. This clearly suggests that the temperature of the earth during much of this period should have been at the freezing level. However, the fossil record indicates that the temperature of the earth during this period remained at levels suitable for life. The hypothesis here is that a vast network of feedback loops between volcanic eruptions, rock weathering, soil bacteria, oceanic algae and the production of limestone sediments maintained earth temperature at this level by regulating the amount of carbon dioxide in the atmosphere.

The new story of biology has also revealed that evolved mechanisms within parts (DNA) manifest as emergent regulatory properties in both organisms and ecosystems. DNA codes for the production of enzymes via a complex network of feedback loops that determine which enzymes are produced. But cell types differ from one another not merely because they contain particular genes. It is also because different sequences in these genes are activated in specific cellular environments, and a protein typically changes its form and function in these environments in a complex network of evolved feedback loops with other proteins.

For example, each of the estimated twenty thousand genes in human DNA codes for the production of a single protein. But this initial input does not determine the function of this protein in a causal, linear fashion. One reason why this is the case is that subsequent interactions with other proteins typically result in smaller sequences that are recombined to form about ten other proteins that perform specialized functions. Another is that a large repertoire of regulatory mechanisms modifies human proteins to perform specialized functions on a moment-to-moment basis in response to a wide range of environmental stimuli.[10]

This explains why interactions between genes at one level during reproduction result in more complex biological systems and processes with distinct functions at another level. For example, feedback loops between proteins, lipids, and nucleic acids result in the emergence of a cell, and other feedback loops result in the emergence of a tissue. This process further unfolds as feedback loops at progressively higher levels of self-organization result in the emergence of organs and entire organisms.

COMPETITION VERSUS COOPERATION

Based on the assumptions that individual organisms, like classical atoms, are atomized, and that the dynamics of evolution acted between or outside organisms, Darwin claimed that competition for survival between organisms was "the rule of nature." He also claimed that this competition is most severe between members of the same species: "The struggle will almost invariably be most severe between the individuals of the same species, for they frequent the same districts, require the same food, and are exposed to the same dangers."[11]

In Darwin's view, the struggle for survival between organisms was the only limiting condition on increases in the population of a species, and in the absence of this struggle, the rate of increase would be exponential: "Every single organic being may be said to be striving to the utmost increase in numbers."[12] And if this "utmost increase" was not checked with competition for survival from other species, the consequences, said Darwin, would be disastrous: "There is no exception to the rule that every organic being naturally increases at so high a rate, that, if not destroyed, the earth would soon be covered by the progeny of a single pair."[13]

Using the example of elephants, Darwin attempted to estimate the rate of increase in the population of this species in the absence of competition with members of other species. He assumed that a pair of elephants begins breeding at age thirty, that the pair continues to breed for ninety years, and that six young elephants are born during this period. If each offspring survives for one hundred years and continues to breed at the same rate, Darwin calculated that nineteen million elephants descended from the first pair would be alive after a period of 740 to 750 years.[14] He then concludes that this natural tendency for species to increase in number without limit is checked by four "external" causes: predation, starvation, severities of climate, and disease.[15]

But we now know that "internal" mechanisms regulate population growth in large numbers of species. Take the example of Darwin's elephants. In a study of over three thousand elephants in Kenya and Tanzania from 1966 to 1968, biologist Richard Laws found that "the age of sexual maturity in elephants was very plastic and was deferred in unfavorable situations." Depending on those situations, individual elephants reached "sexual maturity at from 8 to 30 years."[16] Laws also found that females do not continue to bear offspring until ninety, as Darwin supposed, but cease to become pregnant around fifty-five years of age. The primary mechanism that regulates the population of elephants is the "internal" adjustment of the onset of maturity in females which lowers the birthrate when overcrowding occurs.

Other studies have shown that internal adjustments in the onset of maturity in females regulate population growth in a large number of species. Linkage between age of first production of offspring and population density has been found in white-tailed deer, elk, bison, moose, bighorn sheep, ibex, wildebeest, Himalayan tahr, hippopotamus, lion, grizzly bear, harp seals, southern elephant seals, spotted porpoise, stripped dolphin, blue whale, and sperm whale.[17] A large number of animal species also internally regulate populations by varying their litter and clutch size in response to the amount of food available.[18]

In the new story of biology, there are also mechanisms within organisms that reduce competition for survival in particular ecological niches and sustain biological diversity. This explains why organisms in the same habitat often display adaptive behavior that results in the division of the habitat into niches where the presence of one species does not compromise the existence of another, similar species. For example, zebra, wildebeest, and gazelle are common prey to five carnivores—lion, leopard, cheetah, hyena, and wild dog. But these predators coexist because feedback loops in the process of evolution have resulted in five different ways of living off the three prey species that do not directly compete with one another.

As biologist James Gould explains: "Carnivores avoid competing by hunting primarily in different places at different times, and by using different techniques to capture different segments of the prey population. Cheetahs are unique in their high-speed chase strategy, but as a consequence must specialize on small gazelle. Only the leopard uses an ambush strategy, which seems to play no favorites in the prey it chooses. Hyenas and wild dogs are similar, but hunt at different times. And the lion exploits the brute-force niche, depending alternately on short, powerful rushes and strong-arm robbery."[19]

Herbivores also display evolved behavior that minimizes competition for scarce resources. Paul Colinvaux says the following about this behavior in the African Savanna: "Zebras take the long dry stems of grasses for which their horsy incisor teeth are nicely suited. Wildebeest take the side-shoot grasses, gathering with their tongues in the bovine way and tearing off the food against their single set of incisors. Thompson's gazelles graze where others have been before, picking out ground-hugging plants and other tidbits that the feeding methods of the others have overlooked and left in view. Although these and other big game animals wander over the same patches of country, they clearly avoid competition by specializing in the kinds of food energy they take."[20]

Also consider that three species of yellow weaver birds in Central Africa live on the same shore of a lake without struggle because one species eats only hard black seeds, another soft green seeds, and the third only insects.[21] In North America, twenty different insects feed on the same white pine, but five eat only foliage, three live off the droppings of birds, three on twigs, two on wood, two on roots, one on bark, and four on cambium.[22] A newly hatched garter snake pursues worm scent over cricket scent, and a newly hatched green snake in the same environment displays the opposite preference. Yet both species of snake could in theory at least eat the same prey.[23]

Evolved mechanisms within parts (species) that manifest as emergent regulatory properties in wholes (ecosystems) are particularly obvious in plants. Each plant in the same environment typically specializes in distinct niches: some thrive in sandy soils, others in alkaline, and some, such as lichens, require no soil. Some grow early in the season and others late, and some get by by being small and others by being huge. In studies of two species of clover in the same field, one grew faster and reached a peak of leaf density sooner, and the other grew longer petioles and higher leaves that allowed it to overtop the faster growing species and avoid being shaded out.[24] And these are just a few of the myriad number of examples from the new story of biology which suggest that cooperation is just as pervasive and could well be more important than competition in the evolution of life on earth.

PARTS AND WHOLES IN PUBLIC POLICY

Organisms are still compared in biology textbooks with factories or machines, and this mechanistic view is pervasive in public debates about the environmental crisis and descriptions of environmental problems in both print and electronic media. But a machine is a unity of order and not of substance, and the order that exists in a machine exists outside or between the parts. Machines are constructed from without, the whole is the assemblage of constituent parts, and the interactions between the parts define the function of the whole. Parts of machines can be separated and reassembled and the machine will run normally. But if we separate a living organism into its component parts, the emergent properties of life vanish.[25] Our fondness for mechanistic explanations also explains why models of DNA resemble futuristic tinker toy machines and wrongly suggest that the building blocks of life exist in fixed and static relationship to one another in a rigid spiral staircase. In reality, the base pairs in this molecule are always moving and vibrating, electrons are constantly migrating, and nothing remains the same for more than for a few milliseconds.[26]

If life could be understood as system in which distinct and separate atomized organisms interact in accordance with deterministic natural laws, we could reasonably expect that the environmental impacts of human activities could be predicted with a high degree of certainty. But this expectation is not reasonable or even rational because life is a nonlinear system that sustains conditions that perpetuate itself via a vast web of evolved and indeterminate feedback loops. Enhanced climate models running on more powerful computer systems may have allowed us to make predictions about the environmental impacts of human activities within an increasingly narrow range of statistical probabilities. But these predictions will never be accurate to a degree demanded by those who believe we live in a Newtonian universe.

In public debates about global warming, this is a large and menacing problem. When, for example, scientists testify before committees in the U.S. Congress and describe the projected impacts of global warming in terms of a range of statistical probabilities, this testimony is almost invariable dismissed by one or more committee members on the grounds that the scientists do not really know what they are talking about. In the view of these committee members, "good science" results in accurate predictions and no action should be taken to deal with the problem of global warming in the absence of these predictions.

LESSONS FROM THE NEW STORY

From the perspective of the new story of science, assumptions about the relationships between parts and whole in the old stories about political and economic life are utterly absurd and dangerously wrong. And this is quite apparent in the map space of planet earth that is grafted onto the surface of the conventional globes that sit in classrooms, libraries, and government office buildings. The parts (sovereign nation-states) on these globes are surrounded by dark lines and painted primary colors, and they exist in fixed and static relationship to the whole (geopolitical reality). The only source of political power in this whole is each part, and the system of international government that is charged with the task of preserving and protecting the environmental resources that sustain human life (United Nations) does not in itself have any political power.

In the narrative that serves as the basis for coordinating economic activities in the geopolitical reality imaged on conventional globes, the parts (national economies) are separate, distinct, and isolated from other parts and from the natural environment that exists outside the parts. In this narrative, the baseline measure of the health of the parts is a three percent increase per annum in Gross National Product (GNP), the whole (global market system) must perpetually grow and expand, and there are no biophysical limits on this growth and expansion. The only natural resources that have economic value are those that are bought, sold, or traded in the production, distribution, and exchange of goods, commodities, and services. And other natural resources, such as potable water, breathable air, biological diversity, sustainable ecosystems, and the global climate system, have no economic value.

The fiction that the map space of geopolitical reality imaged on conventional globes of the world actually exists in the real world is perpetuated by narratives about nationalism and national identity. In these narratives, the sameness of people living within the artificial boundaries of a particular part (sovereign nation-state) is defined in terms of differences from people living within the borders of other parts. Those who assimilate these narratives are led to believe that their primary obligation is to protect and defend the vested interests of a particular sovereign nation-state and the territory that lies within its borders. And the parts endlessly compete with one another to secure the natural resources needed to fuel the growth and expansion of their national economies and the capital resources required to increase their political influence and military might.

The new story about the relationships between parts and whole in biological reality is imaged in the digital photographs and videos beamed down from the international space station and earth observing satellites. In this real world, parts (organisms) are emergent from and embedded in the whole (system of life), and the web of interactions between the parts sustains conditions that perpetuate the existence of this whole. Everything in this world is quite literally connected with everything else, and human and environmental systems are embedded in and interactive with one another on the local, regional, and global levels.

The new story of science has also revealed that the part in this real world that we call self is emergent from and embedded in the seamlessly integrated whole of the cosmos on the micro or quantum level and the seamlessly interconnected whole of the system of life on the macro level. Self in this story is not separate from world, mind and nature do not exist in disparate domains of reality, and there is no basis for assuming that we are alone in an alien universe or trapped, as Nietzsche put it, in "a prison house of language."

There is another implication of the new story of science that could be vitally important in the effort to resolve the environmental crisis that is consistent with our scientific worldview but not subject to proof in scientific terms. If members of our species emerged during the later stages of the evolution of the seamlessly integrated whole of the cosmos as fully conscious beings, and if this cosmos was evolving toward higher levels of complexities in the early stages, we can logically conclude, as opposed to scientifically prove, that the universe is conscious.[27] I do not mean to suggest that this consciousness is in any sense anthropomorphic or that this view of progressive order in the evolution of the cosmos legitimates any conceptions of design, meaning, purpose, intent, or plan in any religious tradition. As Heisenberg put it, such words are "taken from the realm of human experience" and are "metaphors at best."[28]

What I mean by conscious universe is, however, consistent with the totality of scientific knowledge and is anthropocentric only to the extent that it answers to a very basic human need. The need is to feel a profound sense of communion with a single significant whole in which the sense of self as a discrete and separate atomized entity ceases to exist. And it is no accident in my view that this sense of communion with a single significant whole is recognized in all of the great religions of the world as the most profound religious experience.

The next four chapters will examine the origins and history of two constructs that are foundational to the old stories about political and economic reality—the sovereign nation-state and the invisible hand. One aim in these chapters is to explain why the resolution of the environmental crisis will require a massive transformation of our political and economic institutions and new standards for moral and ethical behavior. The other is to make the case that these remarkable developments could occur if sufficient numbers of spiritually aware and environmentally concerned people enter the new dialogue between the truths of science and religion.

CHAPTER 5

The Old Story: Sovereign Nation-States and Myths of National Identity

The fifteenth and sixteenth centuries are celebrated for the voyages of discovery that proved the world is round. The eighteenth century saw the first proclamations of universal human rights. The twentieth century's conquest of space made it possible for a human being to look at our planet from a point of view not on it, and to see it, literally, as one world. Now the twenty-first century faces the task of developing a suitable form of government for that single world. It is a daunting moral and intellectual challenge, but one we cannot refuse to take up. The future of the world depends on how well we meet it.

Peter Singer

When members of a society coordinate their activities based on a broadly disseminated and reinforced set of dogmatic beliefs in their mythological or religious traditions, anthropologists refer to these beliefs as useful myths. The aim of this chapter is to reveal that the dogmatic beliefs associated with the construct of the sovereign nation-state are useful myths that can no longer be viewed as useful because they are effectively undermining efforts to resolve the environmental crisis. This situation is greatly complicated by the fact that the sovereign nation-state is a normative construct, or a construct that is assumed to be a taken-for-granted and indelible aspect of geopolitical reality. The large problem here is that this normative construct constitutes one of the greatest conceptual barriers to resolving the environment crisis.

This brief account of the origins and transformations of the construct of the sovereign nation-state is intended to accomplish four objectives. The first is to demonstrate that the construct of the sovereign nation-state emerged in Europe from the eleventh to the sixteenth centuries in a series of narratives that transferred the God-given power and authority of sovereign monarchs to the states governed by these monarchs. The second is to reveal that the narratives about nationalism and national identity that emerged during and after the Protestant Reformation abused the truths of religion in an effort to convince core populations living within the borders of particular nation-states that they were a chosen people possessing superior cultural values and personal qualities. The third is to show that the dogmatic beliefs legitimated and perpetuated by these narratives eventually resulted in the creation of churches of state with sacred symbols, rites, and rituals similar to those in Protestant and Catholic churches. And the fourth objective is to provide a basis for understanding how these dogmatic beliefs eventually became foundational to a system of international government, the United Nations, predicated on the construct of the sovereign nation-state.

The history of this construct is much more complex and far more detailed than the brief account in this chapter suggests. But it does provide the background required to recognize and come to terms with two very inconvenient truths. The first is that it is not possible in a system of international government in which the only source of political power is the sovereign nation-state to implement the scientifically viable public policies and economic programs required to resolve the environmental crisis. And the second is that we must begin very soon to replace the existing system of international government with a supranational federal system capable of implementing these public policies and economic programs.

ORIGINS OF THE CONSTRUCT OF THE SOVEREIGN NATION-STATE

The history of the construct of the sovereign nation-state begins in the eleventh century in Europe when people lived in isolated towns and villages and the religious narrative taught by the Catholic Church made it possible to identify in meaningful ways with other groups in Christendom. States in this century were kingdoms, and kings were viewed, in accordance with pre-Christian beliefs, as hybrid or mixed personalities with mystical or supernatural powers. And the status of kings as both monarch and high priest was confirmed during coronations by anointment with holy oil.

Prior to the thirteenth century, the office of the king was viewed as inseparable from the person of the king, the death of a king signaled the demise of his office, and the vassals were released from their bond of fealty. When a king died, the fief governed by this overlord typically collapsed, vassals battled with one another to establish control over the lands in this fief, and the winner of this competition typically became the new overlord or king.[1] In an effort to prevent this breakdown in the societal order, new narratives were created that enlarged the bases for believing in the divine rights of kingship while simultaneously legitimating the idea that the person of the king could be divorced from the office of the king.

The most influential of these narratives, written by Thomas Aquinas in the thirteenth century, appealed to the work of Aristotle and Augustine to make the case that the secular authority of the king and the office of the king were disparate but related aspects of God's ordered universe. And this narrative was widely used in the late thirteenth and fourteenth centuries to legitimate the idea that absolute monarchies were secular institutions invested with their own authority in a domain of reality distinct from but intimately connected with the spiritual authority of the Catholic Church.

During this period the papacy was the ultimate authority in spiritual affairs and the final arbiter of religious truths. But since the papacy was also actively engaged in protecting its interests in both financial and political terms, the conflict with the states governed by monarchs increased during the fifteenth and sixteenth centuries to crisis proportions. Efforts to mitigate this conflict resulted in new narratives that transferred beliefs and sentiments associated with the shared linguistic and cultural heritage of core populations to the construct of the state by embedding this construct in revisionist histories infused with religious ideals.

In these narratives, people in these core populations were depicted as part of a unique "super-family" that possessed what Weber called "irreplaceable cultural values."[2] Most of these narratives described a golden age in the history of a core population that was closely modeled on the biblical story of the liberation of the Israelites from Egypt by Moses and the subsequent creation of the kingdom of David. And one of the major themes in these tales of struggle and redemption was that God would confer special favor on members of this super-family and preserve and protect their sacred territories or promised lands in exchange for the faithful observance of moral standards and legal codes.[3]

The narrative that redefined the state as a secular institution which exists in a domain of reality separate from spiritual reality was created by the Italian Niccolò Machiavelli in the late fifteenth century. This sometimes chancellor of the Florentine Council claimed that "all men are evil and constantly give way to their evil inclinations the moment they have the chance."[4] In *The Prince*, Machiavelli argued that the only defense against this evil was the state. And the state as Machiavelli defined it was a self-justifying form of government in which political authority is uniformly exercised irrespective of the person or persons who exercise this authority.

According to Machiavelli, the state can exercise its powers without appeal to moral universals, the aim of the state is to maintain itself at all costs and with every possible means, and the job of the prince is to protect and enhance the power of the state. Machiavelli also claimed that the rise and fall of states is governed by laws separate and distinct from those of religion and that great statesmen operate in accordance with their own rules and codes of behavior. Even more remarkable, he exhorted his readers to care more about the future of the state than the fate of their immortal souls.

In the sixteenth century, loyalty to the Church was often in direct conflict with loyalty to the state, and monarchs appealed to Machiavelli's narrative to enhance belief in their absolute authority and centralized governments. As sociologist Arthur Stinchcombe puts it, loyalty to the state seeks "to suppress internal divisions within the nation and to define people outside the group as untrustworthy as allies and implacably evil as enemies.... It is on the one hand a generous spirit of identification...a love of compatriots.... But it is on the other hand a spirit of distrust of the potential treason of any opposition within the group and a hatred of strangers."[5] Studies on the psychological dynamics of this process have revealed that loyalty to an in-group can be solidified by discrimination against an out-group.[6] The narratives in fifteenth and sixteenth century Europe that exploited these dynamics enhanced loyalty between people in the core linguistic and cultural communities of nation-states by appealing to what Freud called the "narcissism of minor difference."[7]

THE EMERGENCE OF CHURCHES OF STATE

After the absolute monarchs of Spain, Portugal, England, and France began to compete with one another in the process of creating vast new empires in the Americas, they soon realized that the winners of this competition must be able to more freely exercise their sovereign powers. They also realized that this was not possible unless there was less interference from the Vatican in the affairs of state and a more profound sense of identification with and loyalty to the state. The narratives that served this purpose began to emerge during the Protestant Reformation in the sixteenth century, and they eventually resulted in the creation of state-sponsored or state-controlled religious systems.

The narrative that triggered the Reformation was authored by Martin Luther, a German monk who studied law before receiving a doctorate in theology and becoming a professor of theology at the University of Wittenberg. The document that marked the beginning of a prolonged period of political conflicts, insurrections, and wars that culminated in a peace agreement predicated on the construct of the sovereign nation-state was Luther's *Ninety Five Theses*. When Luther nailed this document on a door of the church at Wittenberg in 1517, he did not intend to foment a large-scale rebellion against the authority of the Catholic Church. His intent was to convince the Vatican to put an end to the practice of selling indulgences that would allegedly provide remissions of the penalties of sins and reduce the time spent in purgatory. But after Pope Leo X concluded that Luther was "some drunken German who will amend his ways when he sobers up" and failed to take the matter seriously, the *Theses* were translated from the original Latin into German, and thousands of printed copies were distributed throughout Germany.[8]

During the political turmoil that followed, Luther advocated a more definite break with the Catholic Church in a political tract that appealed to the German princes to challenge the power of the papacy by establishing a reformed church in Germany. That same year Luther wrote a pamphlet claiming that the sacramental system of the Catholic Church was the means by which the Vatican had held the real meaning of the Gospel hostage for a thousand years. This document also claimed that salvation is assured through faith in the promises of God signified by the crucifixion of Jesus and that the *Bible* is the sole authority in the pursuit of religious truth. As it turned out, these two ideas, salvation is achieved by faith alone and the Bible is the single source of religious truths, became the twin pillars of the Protestant Reformation in Europe.

In the second generation of Protestant reformers, the author of the narrative that had the largest impacts on the historical developments which resulted in a system of international government predicated on the construct of the sovereign nation-state was John Calvin. Educated in his native France, Calvin was forced to flee to Switzerland after his conversion to Protestantism, and it was there that he published the first edition of his most influential book, *Institutes of the Christian Religion*, in 1536. Calvin adhered to Luther's doctrine of salvation by faith alone but placed more emphasis on the absolute sovereignty of God. Based on the assumption that an all-powerful and all-knowing God freely chooses those who will be saved or damned, Calvin argued that salvation or damnation is predestined by eternal decree. And this became the basis for the Calvinist doctrine that the elect could be recognized by signs of their election, such as material prosperity and impeccable moral behavior.

In Geneva, Calvin displaced the city government with a church government and created a court that had the right to severely punish those who deviated from the teachings of this church. Missionaries trained in Geneva were sent to all parts of Europe, and Calvinism soon became a force to be reckoned with in France, the Netherlands, Scotland, and central and Eastern Europe. According to Calvin's teachings, the devout were morally obligated to engage in acts of resistance against any monarch who persecuted them on account of their religious faith, even if this resistance resulted in armed combat and civil war. The French Calvinists, known at the Huguenots, acted in accordance with this dictum and in 1562 and instigated a civil war that for the next thirty-seven years embroiled the entire country in a series of bloody conflicts.

During the Reformation, the states of Europe were almost constantly at war. The first battle in the most destructive of these wars began in 1618 in the Germanic lands of the Holy Roman Empire between Catholic forces led by the Habsburg Holy Roman emperors and Protestant forces under the command of nobles in Bohemia. The conflict soon widened to include other states on both sides of the religious and political divide, such as Denmark, Sweden, France, and Spain.

Bloody wars between these states continued for thirty years, and the agreement that finally ended this bloodshed, the Peace of

Westphalia, was signed in 1648. This agreement stipulated that signatory states were free to determine their own religion, that the roughly three hundred states that made up the Holy Roman Empire were to be regarded as sovereign nation-states, and that future diplomatic relationships and international agreements would be predicated on this conception of the state. For our purposes, this is a seminally important document in human history for two reasons. First, the Peace of Westphalia transferred metaphysical assumptions about the God-given power and authority of absolute monarchs to the construct of the sovereign nation state. And second, these assumptions remained essentially unchanged, which is why the only source of political power in the present system of international government, the United Nations, is the sovereign nation-state.

NARRATIVES ABOUT NATIONALISM AND NATIONAL IDENTITY

The intent of the new narratives about nationalism and national identity that emerged from the late seventeenth through the twentieth century was to inculcate a profound sense of identification with and loyalty to sovereign nation-states. Those involved in the process of creating these narratives realized that a common language is a critically important tool in achieving this objective. Protestant countries had an initial advantage because during the Reformation the Bible in these countries had been translated into a standard national language, and all official business of state-sponsored or state-controlled Protestant churches, including church services, was conducted in those languages. This explains why the Saxon-Meissen dialect used by Luther in his translation of the Bible quickly became the language of the educated classes in German-speaking regions and why the Hussite Bible, printed in the local dialect of Prague, contributed to the standardization of the Czech language.

The new narratives about nationalism and national identity depicted the members of core populations in sovereign nation-states as the direct descendents of a virtuous and heroic people who lived in the service of irreplaceable cultural values in a mythical golden age. For example, in the eighteenth century the brothers Grimm published tales that were allegedly faithful accounts of primeval Germanic myths which celebrated the more admirable qualities of the German people. The brothers claimed that these myths were verbatim accounts gathered from peasants in the Upper Rhine.

But actually they were drawn from earlier collections, and most of their folksy charm was supplied by the editors. During the decades prior to World War II, lyric poets and popular novelists in Germany painted romantic pictures of a mythical age in which people in the Holy Roman Empire lived without conflict or confusion and faithfully practiced Christian piety. In these narratives, evil was attributed to two external influences—the invidious manipulations of the Vatican or the utter duplicity of the French.

Many narratives about nationalism and national identity in the nineteenth century took the form of revisionist histories about core populations in sovereign nation-states that were more fiction than fact. And these histories were often included in textbooks used by students in state-sponsored and -controlled centralized educational systems. The focus in these textbook accounts, as Hagen Schulze puts it, was on the "new idea of the one, indivisible and immutable nation, born from the ancient spirit of its people." And the intent of these revisionist histories was to convince students that their lives were intimately connected with "a coherent and seamless past from which might be deduced the fateful continuity that justified the nation for all time."[9]

For example, Heinrich Luden's twelve-volume *History of the German People*, which began to appear in 1925, depicted a halcyon age in which members of Germanic tribes won a seemingly endless series of heroic battles and acted in accordance with beliefs and values that were extremely admirable and quintessentially German.[10] In spite of the fact that Luden did not pay much heed to historical evidence and made frequent references to blatantly unscientific racial theories that were later embraced by Adolf Hitler, his version of the history of the German people became the standard version.

What is most important to realize here is that the narratives about nationalism and national identity conflated the truths of religion and the construct of the nation-state to the point where religion became another name for the state.[11] The emotional appeal of these narratives and the widespread acceptance of this view of the nation-state resulted in the emergence of churches of state. Ritualized acts of worship were organized around symbols of nationhood, such as flags, anthems, and military parades; the heroes of the state were honored with place-names and statues; and the accomplishments of these heroes were memorialized in stories, myths, and legends.

During the 1930s, the more invidious aspects of churches of state became painfully apparent in the mass political rallies and military parades held in Germany, Italy, and the Soviet Union. And the willingness of the German people to embrace the assertion by the high priest of the Nazi church of state that his vision of the future of the Third Reich was sacredly ordained was largely responsible for the acts of aggression on neighboring states that led to World War II. It also is reasonable to assume that the rites and rituals of churches of state in the Allied countries created a political climate in which the morally bankrupt decision was made to drop napalm bombs on civilian populations in Germany and atomic bombs on civilian populations in Japan.

THE PRESENT SYSTEM OF INTERNATIONAL GOVERNMENT

The legacy of this past is present in the organization and decision-making processes of the system of international government created after World War II—the United Nations. This system consists of a Security Council that has decision-making authority and a General Assembly that does not. The five permanent members of the Security Council represent the major powers that were victorious in World War II, and each has absolute veto power over any actions taken by this body.

The General Assembly, which is composed of representatives from 189 sovereign nation-states, elects ten additional members of the Security Council for two-year terms. But no substantive decisions can be made by the Security Council if there is any overt opposition from the five permanent members. And since the founding documents of the United Nations stipulate that the General Assembly "can take action only in very limited circumstances," the appearance of egalitarianism is misleading.[12] In this body, the government of India has the same voting power as the government of Ireland, and the states with the smallest populations can easily defeat a resolution proposed by those with the largest populations.

Numerous international agreements have been implemented under the auspices of the United Nations that were intended to resolve problems in the global environment. But a sovereign nation-state has never endorsed an agreement that privileges the goal of achieving a sustainable global environment over its own perceived vested interests, and it is easy to understand why this is the case. If a sovereign nation-state elected to take this unprecedented step, this would create a political power vacuum and severely compromise the ability of this state to defend its vested interests. The unfortunate result is that most proposals that could potentially reduce the destructive environmental impacts of human activities are never formally considered. If a proposal survives the negotiation process, the final agreement almost invariably makes a mockery of the original intent by failing to implement the scientifically viable solutions required to resolve an environmental problem.

For example, Norway, Japan, and Greece, which have large shipping industries, blocked agreements on marine pollution from oil tankers. But Germany, Italy, the Netherlands, and Sweden were more flexible in these negotiations because their economies are much less dependent on these industries. Norway violated an agreement it had previously signed that was intended to prevent the extinction of whales and defended this action before the international community.[13] The United States, in contrast, had no difficulty taking the leading role in the passage of this agreement because it no longer had a whaling industry and did not wish to offend the growing numbers of Americans associated with the save-the-whales movement.

Thirty-two small island nation-states, along with those with densely populated coastal plains, such as Bangladesh, Egypt, and the Netherlands, actively supported initiatives to reduce carbon dioxide emissions to curb global warming because rising sea levels will imperil the existence of a significant percentage of their populations. But the representatives of economically prosperous industrialized states massively resisted these initiatives because they were convinced that the reductions in emissions would retard the growth and expansion of their economies.

In the standard model for forging agreements under the auspices of the United Nations that could resolve environmental problems, the first step is to negotiate a general framework convention. This document defines the environmental problem and the broad policy issues involved in the process of attempting to resolve this problem. If the negotiations do not break down at this stage, the framework convention could be implemented in a regime, an evolving system that defines the problem in more specific terms, the action-oriented protocols that could solve the problem, and the procedures and rules that should be followed. One of the major reasons why the agreements that survive this process have been hugely ineffectual is that the legal principle of state sovereignty allows governments to protect their perceived vested interests during every stage of the negotiations.

This explains why the Framework Convention on Climate Change (1992) failed to protect the climate system, the Convention on Biological Diversity (1992) did not even begin to reduce losses in biodiversity, and the Convention to Combat Desertification (1994) did not slow, much less reverse, this process. The Convention on the Law of the Sea (1982) and a host of other international agreements intended to reduce ocean pollution, prevent over-fishing, and protect endangered species failed to meet any of these objectives. Nonbinding principles that could serve as the basis for the sustainable management of forests were agreed to at the Earth Summit (1992), but negotiations broke down prior to the point where a general framework convention could be articulated. A Convention on the Non-Navigable Uses of International Watercourses has been negotiated, but it has not gone into effect because some sovereign nation-states perceive this agreement as a threat to their national interests.[14]

Scientific evidence may play a supportive and enabling role in some negotiations but only as a minimum condition for serious consideration of an environmental issue. For example, numerous scientific studies on the damage done to European forests by sulfur dioxide emissions led to an agreement in 1985 that reduced these emissions to thirty percent of 1980 levels. Similarly, the scientific evidence presented in the Second Assessment Report of the Intergovernmental Panel on Climate Change (IPCC) was partially responsible for the passage in 1997 of the Kyoto Protocol to the Framework Convention on Climate Change.

But what is not widely known is that these agreements made a mockery of the scientifically based solutions. In the vast majority of negotiations on a great range of issues, such as commercial whaling, trading hazardous waste, loss of biodiversity, conditions in the Antarctic, and ocean dumping of radioactive waste, the scientific evidence was not given serious consideration. When this evidence was perceived as a direct threat to the perceived economic interests of particular sovereign nation-states, it was either systematically ignored or explicitly rejected by the representatives of these states.[15]

THE LEGAL PRINCIPLE OF STATE SOVEREIGNTY

Perhaps the best way to illustrate how appeals to the legal principle of state sovereignty effectively undermine the prospect of implementing scientifically viable solutions for environmental problems is to examine the role played by American diplomats in the process of forging the terms of the post-Kyoto agreement on global warming. During the meeting in Brussels in 2007, diplomats from more than one hundred forty countries were discussing the language used in a document called the Summary for Policy Makers. This document summarized a compendium of scientific research on global warming compiled over a six-year period by twenty-five hundred scientists associated with the United Nations–sponsored IPCC.

The IPCC scientists who participated in the meeting in Brussels argued that the language in the Summary for Policy Makers was entirely consistent with the results of the scientific research and should not be changed. But the American diplomats representing the administration of President George W. Bush tried to eliminate or change any statements in the report that they perceived as a potential threat to American economic interests and to the growth and expansion of the global market system.[16] These diplomats attempted over a hundred times to change unambiguous phrases describing the projected impacts of global warming, such as "will happen," to such ambiguous phrases as "will likely happen." They also tried to eliminate a section of the report that would require highly industrialized countries in the Group of Eight to reduce worldwide greenhouse gas emissions to fifty percent below 1990 levels by 2050. The members of the U.S. delegation even tried to delete the following key statement: "We firmly agree that resolute and concerted action is urgently needed in order to reduce global greenhouse gas emissions and sustain our common basis of living."[17]

During the meeting in Bali in December 2007, the American diplomats refused to consider the proposal in the final draft of the Summary for Policy Makers that industrialized countries should reduce their emissions of greenhouse gases to forty percent below 1990 levels by 2020 and fifty percent below these levels by 2050.[18] After this proposal was removed from the document's binding targets section and placed as a footnote in the nonbinding preamble section, the American delegation tried to block a proposal that industrialized countries should provide poor countries with financial aid and technological assistance. At the point at which it appeared that the meeting would conclude without any agreements on this issue, the head of the U.S. delegation, Under Secretary of State Paula Dobrianski, withdrew the objection to the proposal.[19] As an editorial writer in *The New York Times* put it, "From the United States the delegates got nothing, except a promise to participate in forthcoming negotiations. Even prying that out of the Bush administration required enormous effort."[20]

When President Bush announced that the United States was unwilling to meet the modest reductions in greenhouse gas emissions provided in the Kyoto Treaty, representatives of the European Union were outraged. During an emergency visit to Washington a few days later, the EU representatives made their case to Christine Todd Whitman, administrator of the Environmental Protection Agency. Whitman told the representatives she was as "optimistic as the President that, working constructively with our friends and allies through international processes, we can develop technologies, market-based incentives and other innovative approaches to global climate change."[21]

The U.S. decision to withdraw from the Kyoto Treaty was also a central point of contention in a meeting between President Bush and the chancellor of Germany, Gerhard Schröder, held a day earlier. When asked to comment on the meeting, Bush said, "We will not do anything that harms our economy, because first things first are the people who live in America."[22] This comment provoked an interesting response from the president of the EU Commission, former Italian Prime Minister Romano Prodi: "If one wants to be a world leader, one must know how to look after the entire earth and not only American industry."[23]

It is worth noting also that virtually every member of the Democratic Party in the U.S. Senate voted against a motion to even consider ratification of the Kyoto Treaty. The members of this party also failed to pass a comprehensive climate bill prior to the 2010 elections, even though they controlled both houses of Congress. Also, President Obama did not provide the leadership needed to secure the votes required to pass this bill, and there are no indications at this point in time that his administration is prepared or willing to deal with the very menacing problem of global warming. And for reasons that will be discussed in some detail later, this was painfully apparent during the last failed attempt by the international community to implement a post-Kyoto agreement on global warming in Durbin, South Africa, in December 2011.

IN FALSE GODS WE TRUST

Philosopher Peter Singer makes a convincing case that cultural evolution resulted in an "expanding moral circle" that moved outward over the course of human history form family, clan, tribe, state, and nation-state. He claims that the expanding moral circle was responsible for the abolition of despotism, slavery, feudalism, racial segregation, and for ongoing attempts to secure the rights and guarantee the freedoms of women and other minorities.[24] Singer also believes that this process will soon result in a universal set of moral principles and standards for ethical behavior that could greatly enhance the prospect of resolving the environmental crisis.

But Singer is also convinced that this crisis will not be resolved unless the present system of international government is replaced fairly soon by a supranational federal system. This system would not be predicated on the construct of the sovereign nation-state, and the member states would be part of a unified federal system of democratic government. In this government, the number of representatives of each nation-state would be proportional to the population of that state, and decision-making authority and political power would reside in the will of all of the citizens in the member states.

In the view of the self-proclaimed realists and pragmatists in the political debate about the environmental crisis, Singer's claim that it will not be possible to resolve this crisis unless the present system of international government is replaced by a supranational federal system is abundantly absurd. But what these critics have apparently failed to realize is that the present system of international government is premised on dogmatic beliefs associated with the construct of the sovereign nation-state that are no longer commensurate with the terms of human survival. And they also seem to be unaware of the fact that these beliefs explain why it is not possible in this system of government to implement scientifically viable solutions for problems in the global environment.

Granted, it is unrealistic to assume that a supranational system of federal government could emerge in the present geopolitical climate. The prediction here, however, is that the escalating impacts of global warming will change this climate in ways that will greatly enhance the prospects that this remarkable development could occur. Research on how massive changes occur in human societies based on dynamical systems theory has shown that during periods of gradual change, negative feedbacks maintain political and economic systems in relative states of equilibrium and radical public policies and economic programs are not given any serious consideration.

But this research has also revealed that during periods of rapid change, a crisis results in positive feedbacks that move political and economic systems toward states that are far from equilibrium, and these feedbacks increase in almost direct proportion to the numbers of people who are aware of and concerned about the crisis. When this occurs in democratic societies, large numbers of previously apathetic citizens become more involved in the political process, many new public policies and economic programs policies are implemented, and fundamental changes in political and economic reality tend to occur over a relatively short period of time.[25]

This means in my view that there will be a window of opportunity in which it will be possible to replace the current system of international government with a supranational federal system. But as Peter Singer has also convincingly argued, it will not be possible to create this system if we fail to extend the moral circle to all of humanity and embrace more universal standards for ethical behavior. It also seems clear that this system will not emerge if we fail to challenge and effectively undermine the assumption that the sovereign nation-state is an indelible aspect of geopolitical reality and the dogmatic beliefs associated with this construct. But it is very unlikely in my view that these massive changes will occur in the absence of a well-organized and highly effective worldwide movement in religious environmentalism.

The next three chapters will examine the origins and transformations of the construct of the invisible hand in mainstream economic theory. This examination will reveal that this construct, like that of the sovereign nation-state, is also a product of and deeply embedded in the Western metaphysical tradition. It will also demonstrate that the economic theory used by virtually all mainstream economists, neoclassical economics, is predicated on metaphysical assumptions about the dynamics of market systems associated with the invisible hand.

One large problem here is that these unscientific assumptions effectively preclude the prospect of implementing scientifically viable economic solutions for environmental problems. The other is that in neoclassical economic theory the magical machinations of the invisible hand function in the minds of the true believers as a religion that features a comprehensive business theology and what theologians call an eschatology, or teachings about the end of history.

One reason why these problems will be extremely difficult to resolve is the widespread belief that neoclassical economics is a rigorously mathematical scientific theory. The discussion that follows will demonstrate that this is not the case. And it will also reveal that the practitioners of this theory are unwittingly worshipping at the altar of a false god and that the price that could soon be paid for this act of devotion is a human tragedy on a scale that even the prophets of the Old Testament would not have dared to imagine.

CHAPTER 6

The Old Story: Metaphysics, Newtonian Physics, and Classical Economics

> It is not from the benevolence of the butcher, the brewer, or the baker, that we expect our dinner, but from their regard to their interest. We address ourselves, not to their humanity but to their self-love, and never talk to them of our own necessities but of their advantages.
>
> Adam Smith

The causes of the environmental crisis may be staggeringly complex, but the most effective way to deal with it in economic terms seems rather obvious. We must begin very soon to implement scientifically viable economic solutions for what is now a large number of very menacing environmental problems. If this could be accomplished within the framework of the theory that now serves as the basis for coordinating global economic activities, neoclassical economics, political leaders, economic planners, and environmental scientists could work together in harmony to implement these solutions. Unfortunately, this cannot happen because neoclassical economic theory is predicated on unscientific assumptions about the dynamics of market systems that effectively preclude the prospect of implementing scientifically viable economic solutions for environmental problems.

These assumptions were articulated in their original form by eighteenth century moral philosophers Adam Smith, Thomas Malthus, and David Ricardo who were members of and greatly influenced by a widespread philosophical and religious movement known as deism.

The fundamental impulse in this movement was to make belief in the existence of the God of the Judeo-Christian tradition consistent with the implications of the mechanistic worldview of Newtonian physics. Because physical laws in this physics completely determine the future state of physical systems, the deists concluded that the universe does not require, or even permit, active intervention by God after the first moment of creation. They then imaged God as a clockmaker and the universe as a clock regulated and maintained after its creation by physical laws.

The moral philosophers we now call classical economists assumed that this deistic god created two sets of laws to govern the workings of the clockwork universe—the laws of Newtonian physics and the natural laws of economics.[1] Based on this assumption, they argued that the forces associated with the natural laws of economics determine the movement and interactions of economic actors in much the same way that forces associated with Newton's laws of gravity determine the movements and interactions of material objects. And this became the basis for the claim that forces associated with the natural laws govern the behavior of economic actors and maintain order and stability in market systems even if the actors are completely unaware that this is the case.

Adam Smith's metaphor for the forces associated with the natural laws of economics was the invisible hand, and this construct became the central legitimating principle in mainstream economic theory. As economists Kenneth Arrow and Hans Hahn put it, the "notion that a social system moved by independent actors in pursuit of different values is consistent with a final coherent state of balance...is surely the most important intellectual contribution that economic thought has made to the general understanding of social processes."[2]

In *The Wealth of Nations*, Smith said the invisible hand is analogous to the invisible force that causes a pendulum to oscillate around its center and move toward equilibrium or a liquid to flow between connecting chambers and find its own level. Based on this analogy, Smith claimed that this unseen hand is the force that moves independent actors in pursuit of different values toward the equalization of rates of return and accounts for the tendency of markets to move from low to high returns. Given that Smith's invisible hand has no physical content and is an emblem for something postulated but completely unproven and unknown, why did he assume it exists? The answer is that Smith was a deist: the invisible hand in his view was that of the absentee deistic god, and his belief in its existence was an article of faith.

Smith, who was notoriously absent-minded, lectured much of his life on problems in moral philosophy at the University of Glasgow. The discipline of moral philosophy was much more broadly defined than it is today and included the study of natural philosophy, ethics, jurisprudence, and political economy.[3] It was on this broad canvas that Smith attempted to depict the inner workings of the mechanistic Newtonian universe and the place of human beings within its "systems, wheels and chains."

Understanding the work of Adam Smith as a whole is recognized as notoriously difficult due to what an earlier generation of German scholars termed "das Adam Smith Problem": that is, the glaring contradiction between the themes in Smith's two books—sympathy in *The Theory of Moral Sentiments* (1759) and self-interest in *The Wealth of Nations* (1776). Some have argued that these contradictions result from some radical change in Smith's worldview in the seventeen-year period between publication dates. But what those who make these arguments do not realize is that Smith consistently and simultaneously revised and edited both books; five editions of each were published in his lifetime, and the final edition of *The Theory of Moral Sentiments* appeared in the year of his death. More important, Smith viewed these books as part of a single corpus, and his reasons for doing so become clear when they are read in this way.

Although the construct of the invisible hand is the ghost in the machine in virtually all of Smith's writings, it is explicitly mentioned only three times in three very different contexts. In the essays which did not appear in his two books, the invisible hand is that of Jupiter, and the construct is used to illustrate how primitive or "savage" people dealt with the irregular phenomena of nature. In *The Theory of Moral Sentiments*, the hand belongs to a deistic providence which ensures that the less fortunate are fed in spite of the greed of the rich. In *Wealth of Nations*, the hand is the metaphor for the natural laws of economics that maintain harmony and stability in market systems with the same impersonal force as the laws of Newtonian physics.

Taken together, the essays and *The Theory of Moral Sentiments* constitute a critique of the history, sociology, and psychology of religion. This critique is designed to demonstrate that during the course of civilization polytheism is replaced by theism and human beings eventually came to realize that nature is a system or machine that obeys both physical and natural laws.[4] In the long essay on astronomy, the invisible hand is that of the god Jupiter and symbolizes the "principles that lead and direct philosophical inquiries," which Smith defines as the passions of wonder, surprise and admiration.[5]

In the section entitled "Of the Origin of Philosophy," Smith claims that polytheism originated among primitive people and was a product of the "vulgar superstition which ascribes all the irregular events of nature to the favour or displeasure of intelligent, though invisible beings, to gods, daemons, witches, genii, fairies."[6] These primitive people, says Smith, were awed by "magnificent" irregularities, such as thunder, lightning, and comets. He then claims that the fundamental problem with the "lowest and most pusillanimous superstition" of these savages is that it prevented them from realizing that "hidden chains" link all events to hidden causes that result from forces associated with deterministic physical and natural laws.

THE SYSTEM OF NATURAL LIBERTY

Smith frequently identifies nature with the way things operate on their own accord and says that the goal of philosophy is to "lay open" the "invisible chains which bind together" the natural world.[7] And the intent of much of his commentary on the system of natural liberty in *Wealth of Nations* is to promote trust in the "natural course of things." This trust is warranted, says Smith, because the "hidden chains" of the invisible hand regulate the "system of natural liberty" and constrict the sphere of human "intention and foresight."[8]

In the beginning of the commentary on the system of natural liberal, Smith says that "no human wisdom or knowledge could ever be sufficient" to provide the sovereign with the ability to effectively manage the "industry of private people" and direct it "toward the employments most suitable to the interests of society." He then argues that since human beings cannot effectively manage market systems or predict their futures, each individual should "pursue his interests in his own way" within "the laws of justice."[9] The usual interpretation of Smith's system of natural liberty is that it legitimates the idea that each of us should have the freedom to pursue our livelihood and self-interest. And since Smith claims that this system can only operate properly if the role of government is limited, it is also widely assumed that he makes government the servant of individualism. The problem with these interpretations is that Smith's system of natural liberty exists in a mechanistic Newtonian universe in which forces associated with physical and natural laws predetermine all events.

This is quite apparent in the section on education in *The Wealth of Nations*. Smith begins by endorsing the idea from the ancient Greeks that philosophy should be divided into natural philosophy, which would later be called physics, moral philosophy, and logic. He then claims that the study of both the human mind and the "Deity" must fall under the province of natural philosophy, which investigates "the origin and the revolutions of the great system of the universe." Smith then concludes that the human mind and the Deity "in whatever their essence might be supposed to consist, are parts of the great system of the universe."[10]

In the physics essay, Smith criticizes those who attempt to explain nature's "seeming incoherence" by appealing "to the arbitrary will of some designing, though invisible beings, who produced it for some private and particular purposes." He then proceeds to fault the superstitious for their inability to conceive of the "idea of a universal mind, of a God of all, who originally formed the whole, and who governs the whole by general laws, directed to the conservation and prosperity of the whole, without regard to that of any private individual."[11] Smith then claims that as our ancestors progressed in knowledge, they realized that nature is "a complete machine...a coherent system, governed by general laws, and directed to general ends, viz., its own preservation and prosperity."[12]

Smith clearly believed that the "universal mind" of God served as a template for the creation of a world that following its creation is entirely governed by general laws. His phrase "general laws" refers to both physical and natural laws, and he implies that they have the same ontological status—both originate from the universal mind of a deistic god, exist in a realm separate and discrete from the material world, and act on atomized parts to maintain the stability of the whole.

The special character of Smith's metaphysics allows us to answer a question that has perplexed most experts on the history of mainstream economic thought: Why does Smith appeal to God to legitimate the existence of the invisible hand in his other major work, *The Theory of Moral Sentiments*, yet refrain from doing so in *The Wealth of Nations*? The answer is that his understanding of the character of natural laws in *The Wealth of Nations* is predicated on the assumption that these laws have co-equal status with physical laws. The omission of any appeals to God not only serves to reinforce the validity of this assumption. It also implies that the real existence of the natural laws is self-evident and any appeal to metaphysics is ad hoc and unnecessary.

THE COSMIC MACHINE AND HIDDEN CHAINS

Smith's metaphysics also allows us to understand why he had no difficulty arriving at the conclusion that natural laws act to preserve human populations without regard for the well-being of individuals. In his view, mind and nature are systems or machines that do not require "personal" intervention because they operate in accordance with forces associated with deterministic laws. This explains why the biblical God, who disrupts the orderly workings of nature by staging miracles and singles out individuals or groups for covenants, revelations, rewards, and punishments, is conspicuously absent in Smith's work. In his view, this God is a product of the frenzied imagination of those who did not realize that nature is "a complete machine" in which "hidden chains" govern the orderly interaction of parts to preserve the existence of the whole.

These ideas are apparent also in Smith's descriptions of the workings of the invisible hand in *The Theory of Moral Sentiments*. Here the hand is part of the "regular" workings of nature and ensures that social benefits result as an unintended outcome of selfish actions. For example, Smith claims that the invisible hand increases the fertility of the earth and benefits the whole of humankind despite inequality in capital resources and ownership of land. But since the landlord, says Smith, can eat only a portion of the produce of his land, the remainder is consumed by those who provide him with luxuries.[13] Hence the rich, despite their "natural selfishness and rapacity," are "led by an invisible hand to make nearly the same distribution of the necessities of life which would have been made had the earth been divided into equal portions among all its inhabitants; and thus without intending it, without knowing it, advance the best interests of society, and afford means to the multiplication of the species."[14]

Smith consistently views ideas as systems and emphasizes the basic similarity between systems and machines. A machine, wrote

Smith, "is a little system created to perform, as well as to connect together, in reality, those different movements and effects which the artist has occasion for." He then says that a system "is an imaginary machine invented to connect together in the fancy those different movements and effects which are already in reality performed."[15] The claim that a machine is a system that performs and connects together parts "in reality" and ideas are a system that describes the connections and interactions of parts "in reality" serves as commentary on Smith's view of economic actors. According to Smith, the natural laws of economics govern decisions made in the realm of ideas by economic actors to maintain the orderly workings of market systems.

If the behavior of economic actors is controlled by the "invisible chains" of natural law, what does this imply about the system of natural liberty? Smith claims that a natural order emerges when conscious interference with an economic system is minimized and that the system of natural liberty contributes to the maintenance of this order. Yet his argument for the existence of the system of natural liberty, as noted earlier, is predicated on the assumption that "no human wisdom or knowledge could ever be sufficient" to create or sustain this order. Some have attempted to resolve this paradox by arguing that Smith distinguishes between a macro-level where natural laws govern the state of market systems and a micro-level where the economic actors enjoy a radical freedom to pursue their self-interests.[16] For Smith, however, the natural laws of economics operate on all levels, and there is no distinction between macro-level order and micro-level actors.

In *The Theory of Moral Sentiments*, Smith says the following about the Stoic philosophers: "A wise man does not look upon himself as a whole, separated and detached from every other part of nature, to be taken care of by itself and for itself." He rather considers himself as "an atom, a particle, of an immense and infinite system, which must and ought to be disposed of according to the convenience of the whole."[17] The argument here is that forces associated with natural laws act on atomized individuals to enhance the welfare of human populations and the freedom of individuals is utterly constrained by these forces. In the "great machine of the universe" with its "secret wheels and springs,"[18] the system of natural liberty may allow the atomized individual to live with the illusion that his or her actions are freely taken. But as the wise man knows, this freedom does not exist because the "connecting chains" of the invisible hand determine decisions made by the parts (atomized economic actors) to sustain the orderly workings the whole (market systems).

THE BUSINESSMAN

This understanding of the machinations of the invisible hand is apparent also in Smith's commentary on the "prudent" businessman in *Wealth of Nations*. The prudent man is praised for his "industry and frugality" and for "steadily sacrificing the ease and enjoyment of the present moment." The reward for this virtuous behavior is that the "situation" of this man grows "better and better every day."[19] In addition to celebrating the economic virtues of the prudent man, Smith also stresses his apolitical character. He claims that the prudent man "has no taste for that foolish importance which many people wish to derive from appearing to have some influence in the management" of public policy. Such a man prefers "that the public business were well managed by some other person" so that he is left to "the undisturbed enjoyment of secure tranquility."[20] The clear inference here is that this tranquility derives from the recognition that order and stability within the larger society results from forces associated with natural laws and that the actions of individuals are governed by these laws.

Smith consistently denigrates the power of human choice and claims that planning, intention, and foresight have had little or no impact on the course of history. He talks a great deal about great moments in history but never suggests that these moments are due to the actions of great men. The explanation for this rather unusual view of human history is that Smith assumes that the actions of individuals are determined by forces associated with the natural laws created by the absentee but benevolent deistic god.

For example, in *The Theory of Moral Sentiments*, at the beginning of the paragraph on the invisible hand, Smith says the following about human history: "And it is well that nature imposes upon us in this manner. It is this deception that rouses and keeps in continual motion the industry of mankind. It is this which first prompted them to cultivate the ground, to build houses, to found cities and commonwealths, and to invent and improve all the sciences and the arts, which ennoble and embellish human life; which have entirely changed the whole face of the globe, have turned the rude forests of nature into agreeable and fertile plains, and made the trackless and barren ocean a new fund of subsistence, and the great high road of communication to the different nations of the world. The earth by these labours of mankind has been obliged to redouble her natural fertility, and to maintain a greater multitude of inhabitants."[21]

According to Smith, the natural laws that govern the movement and interaction of human beings are not dependent in their operation on the intelligence and creativity of individuals, even in the sciences and the arts. Equally significant for our purposes, he also claims that the laws that govern market systems obliged "earth" to "redouble her natural fertility" to maintain the growing human population.

Since Smith favored wage regulation and argued against an inequitable division of wealth, some have argued that his concept of free markets was a liberal proposal to free the poor and the powerless from economic oppression. There is no doubt that Smith was concerned about the plight of the working poor in an era in which conditions of life for this population were quite horrific by modern standards. But in his view, the agency that would improve these conditions is the absentee but benevolent deistic god with the invisible hand. Consider, for example, what Smith says about the problems associated with the inequitable division of wealth: "No society can surely be flourishing and happy, of which the greater part of the members are poor and miserable. It is but equity, besides, that those who feed, clothe, and lodge the whole body of the people, should have the produce of their own labour as to be themselves tolerably well fed, clothed and lodged."[22]

The assumption here is that forces associated with the natural laws of economics are equitable and benefit the "whole body of the people." Since this whole, says Smith, cannot be "flourishing and happy" if the "greater part" is "poor and miserable," the workers should have the benefit of the "produce of their own labor" and be "tolerably well fed, clothed and lodged." Smith then argues that these "tolerable" living conditions for the working poor will enhance efficiency and productivity of the market system, thereby benefiting all members of society.

TIGHTENING THE CHAINS: THOMAS MALTHUS AND DAVID RICARDO

Thomas Malthus and David Ricardo embraced Smith's understanding of the natural laws of economics and attempted to tighten the "invisible chains" that "connect" parts in the machine of the market system in a more rigidly deterministic way. Malthus's *An Essay on the Principle of Population as It Affects the Future Improvement of Society* was written in response to the views of a utopian thinker named William Godwin. In Godwin's account of the human future, the sexual passions that lead to increases in birthrate would somehow diminish after a universal harmony in social and political reality is achieved. When that occurs, wrote Godwin, "there will be no war, no crime, no administration of justice, as it is called, and no government. Besides there will be no disease, anguish, melancholy, or resentment."[23]

Malthus's dissenting views were published anonymously in 1798. The usual interpretation of his principle of population is straightforward—since population increases geometrically and food supply increases arithmetically, there is a tendency for population growth to outrun the means of subsistence. Note, however, the manner in which this argument is actually made:

> "I think I may fairly make two postulata.
>
> First, That food is necessary to the existence of man.
>
> Secondly, That the passion between the sexes is necessary and will remain nearly in its present state.
>
> These two laws, even since we have had any knowledge of mankind, appear to have been fixed laws of nature, and, as we have not hitherto seen any alteration of them, we have no right to conclude that they will ever cease to be what they now are, without an immediate act of power in that Being who first arranged the system of the universe, and for the advantage of his creatures, still executes, according to fixed laws, all its various operations."[24]

Malthus, an ordained clergyman, allows for the prospect that the "laws of nature" could be altered by an "immediate act of power in that Being who first arranged the system of the universe." But this "Being" in his view is the god of the deists. For example, the claim that "fixed natural laws," like the laws of physics, govern the "various operations" of the "system of the universe" in a causal and deterministic fashion clearly implies that the system does not need or require any intervention by God.

While Smith argued that the natural laws that determine the future of markets are essentially benevolent, Malthus concluded that the laws of population could ultimately threaten the existence of humanity:

> "Assuming then, my postulata as granted, I say that the power of population is indefinitely greater than the power in the earth to produce subsistence for man.
>
> Population, when unchecked, increases in a geometrical ratio. Subsistence increases only in an arithmetical ratio. A slight acquaintance with numbers will show the immensity of the first power in comparison of the second.
>
> By that law of our nature which makes food necessary to the life of man, the effects of these two unequal powers must be kept equal.
>
> This implies a strong constantly operating check on population from the difficulty of subsistence. This difficulty must fall some where and must necessarily be severely felt by a large portion of mankind."[25]

Malthus clearly suggests that the unequal powers associated with these natural laws could have disastrous consequences for all of humankind. But he also claims that this will not occur because the interplay between human beings results in "a strong constantly operating check" on population growth that impacts only a "portion" of humankind. Assuming, like Smith, that impersonal and deterministic natural laws govern the interaction between parts (atomized individuals) to perpetuate the existence of wholes (human populations), Malthus concludes that we should not interfere with the operation of the laws. In his view, the deaths of large numbers of the working poor are the unfortunate but inevitable result of the operation of the laws created by the absentee deistic god.

The natural laws of Smith may be more benevolent and less menacing than those of Malthus, but there is no difference between them in ontological terms. In the view of both Smith and Malthus, these laws were created by a deistic god who withdraws from the universe after the first moment of creation, exist in a realm prior to and separate from physical reality, and act causally and deterministically on atomized individuals. The assumption that the "unseen chains" of the laws of population, like those of the natural laws associated with the invisible hand, cannot be broken served to reinforce the view that social–political problems do not lie within the domain of economic theory. And it also allowed subsequent generations of economists to more effectively argue that the business of economists is to describe the lawful workings of free market systems and not to concern themselves with problems that exist in "other" domains of reality.

David Ricardo was the son of Jewish merchant–banker, and his expertise as a stockbroker allowed him to retire at age forty-two with a large fortune. The system or machine of the market in Ricardo's *On the Principles of Political Economy and Taxation* is as abstract and unadorned as a linear equation, and the atomized entities within this system are "forced" to obey the "laws of behavior." Workers appear, as Robert Heilbroner puts it, as "undifferentiated units of economic energy, whose only human aspect is a hopeless addiction to what is euphemistically called 'the delights of domestic society.'" And capitalists are depicted as "a gray and uniform lot, whose entire purpose on earth is to accumulate—that is, to save profits and to reinvest them by hiring more men to work for them; and this they do with unvarying dependability."[26]

One of Ricardo's burning ambitions was to repeal a set of laws passed by a majority of landowners in Parliament that were designed to prevent cheap grain from being imported into Britain. In Ricardo's view, the so-called corn laws were an obvious impediment to improving national welfare, and most of his economic theory is intended to demonstrate that this is the case. Like Smith, Ricardo believed that market systems tend to expand and that the resulting new shops and factories create more demand for labor. As the population increases, the increased demand for grain would, he says, result in higher prices and in the cultivation of more marginal land.

Ricardo argues that while fertile land in earlier times was a gift of nature that existed in such abundance that it was regarded as free, progress has resulted in a situation where fertile land has become scarcer and capital investment is required to increase production. He then claims that the cultivation of more marginal land necessarily increases the overall costs of production and that this is reflected in higher prices for grain and increases in rent for the landlord who owns the best land. Rent as Ricardo defines it is the difference in profits that results when the costs of growing crops on fecund land are less than those of growing crops on less fecund land. In both instances, landlords must pay the same wages and bear the same capital expenses. However, the landlord who owns the more fertile land reaps more profits than his competitors. The problem here in

Ricardo's view is that those who are responsible for this progress, the capitalists, are obliged to pay higher subsistence-level wages to workers while the rising aggregate of rents increases the profits of landlords.

According to Ricardo, the "laws" of supply and demand determine whether quantities in nature are free or have monetary value. When land is plentiful, there is no rent because land has no market value and no one would elect to use his "capital" to buy and cultivate what exists in abundance. The following passage further illustrates this point: "If air, water, the elasticity of steam, and the pressure of the atmosphere, were of various quantities; if they could be appropriated, and each quality existed in only moderate abundance, they, as well as the land, would afford a rent, as the successive qualities were brought into use. With every worse quality employed, the value of the commodities in the manufacture of which they were used would rise because equal quantities of labor would be less productive. Man would do more by the sweat of his brow, and nature perform less; and the land would no longer be imminent for its limited powers."[27]

Based on the assumption that the only natural resources that have value are those that are subject to the law of supply and demand, Ricardo claims that the resources of nature are free unless, or until, they become sufficiently scarce to warrant the capital investment that would allow for their appropriation and sale. Although he implies that more of these resources will have value due to the inevitable expansion of markets and the associated emergence of new manufacturing techniques, the assumption that lies at the core of his argument is that the powers of nature are inexhaustible. It is this assumption that allows Ricardo to argue that there are no limits on the growth and expansion of market systems. And the economic theory now used by virtually all mainstream economists is predicated on this assumption.

THE NOT-SO-WORLDLY PHILOSOPHERS

Some historians of mainstream economic theory make passing mention of the fact that Adam Smith was a deist. But they typically avoid saying anything about the influence of deism on Smith's economic thought by extracting well-known passages from *Wealth of Nations* and treating them as pieces of revisionist history. Others seek to avoid the problem by arguing that *Wealth of Nations*, given the absence of appeals to the Judeo-Christian God, is an entirely secular study of economic reality that is different in kind from Smith's other works. The truth is, however, that Smith firmly believed that the absentee deistic god created the natural laws of economics and that forces associated with these laws determine decisions made by economic actors and maintain order and stability in market systems. If one reads the works of Smith in their entirety, this conclusion is quite impossible to avoid.

What is most remarkable about the role played by metaphysics in the work Adam Smith is that it served to justify the belief that greed is good in a very profound moral sense. According to Smith, decisions made by economic actors in the pursuit of self-interest are governed by forces associated with the natural laws of economics created by an absentee but benevolent deistic god. In his view, the pursuit of self-interest is sanctioned by the providential deity that created these laws and serves the greater good.

That Smith was a deist and that the invisible hand was his metaphor for the forces associated with the natural laws created by a deistic god has largely been forgotten. But what survived is the belief that these laws were created by the God of the Judeo Christian tradition. Later in this discussion it should become clear that this not only explains why the true believers in the benevolent machinations of the invisible hand conflate the natural laws of economists with the laws of this God. It also explains why these true believers assume that market forces that allegedly result from the operations of these natural laws are part of a sacredly ordained providential plan and should not be interfered with by government or any other agency.

The next chapter, on the history of the construct of the invisible hand in mainstream economic theory, seeks to accomplish four objectives. The first is to demonstrate that neoclassical economic theory was created by substituting economic variables derived from classical economics and associated with the invisible hand for physical variables in the equations of a mid–nineteenth century theory in physics. The second is to reveal that the mathematical formalism that resulted from these substitutions was predicated on unscientific assumptions about the lawful dynamics of market systems. The third is to show that these assumptions remained essentially unchanged in spite of the fact that subsequent generations of mainstream economists extended and refined this mathematical formalism. And the fourth is to explain why the unscientific assumptions in neoclassical economic theory and in the mathematical formalism used by mainstream economists effectively preclude the prospect of implementing scientifically viable solutions for environmental problems.

CHAPTER 7

The Old Story: Metaphysics, Mid–Nineteenth Century Physics, and Neoclassical Economics

> The ideas of economists and practical philosophers, both when they are right and when they are wrong, are more powerful than is commonly understood. Indeed, the world is ruled by little else. Practical men, who believe themselves to be quite exempt from any intellectual influence, are usually the slaves of dead economists.
>
> John Kenneth Galbraith

In economics textbooks, the nineteenth century creators of neoclassical economics theory, Stanley Jevons, Léon Walras, Francis Ysidro Edgeworth, and Vilfredo Pareto, are credited with disclosing the dynamics of market systems and transforming the study of economics into a rigorously mathematical scientific discipline. There are, however, no mentions in these textbooks, or in all but a few books on the history of economic thought, of a rather salient fact. Neoclassical economic theory was created by substituting economic constructs derived from classical economics and associated with the invisible hand for physical variables in the equations of a badly conceived and soon-to-be outmoded mid–nineteenth century theory in physics.[1]

The theory in physics that the economists used as the template for their theories was developed from the 1840s to the 1860s. During this period, physicists responded to the inability of classical physics to account for the phenomena of heat, light, and electricity with a profusion of hypotheses about matter and forces. In 1847 Hermann-Ludwig Ferdinand von Helmholtz, one of the best known and most widely respected physicists at this time, posited the existence of a vague and ill-defined energy that could unify these phenomena. This served as a catalyst for a movement called energetics, in which physicists attempted to explain very diverse physical phenomena in terms of a unified and protean field of amorphous energy.

Because the physicists were unable to specify the actual character of this energy and could not be precise about what was being measured, their theories were not subject to repeatable experiments under controlled conditions. The amorphous character of energy in the physical theories also obliged the physicists to appeal to the law of conservation of energy, which states that the sum of kinetic and potential energy in a closed system is conserved. This appeal was necessary because it was the only means of asserting that the vaguely defined system described in the theory somehow remains the "same" as it undergoes changes and transformations.[2]

The strategy used by the creators of neoclassical economics was as simple as it was absurd—they wrote down the equations from the theory in physics that finally emerged in the energetics movement and substituted economic variables for the physical variables. Utility was substituted for energy, the sum of utility for potential energy, and expenditure for kinetic energy. The forces associated with utility–energy were represented as prices, and spatial coordinates described quantities of goods.[3]

Because the physical system described in the equations borrowed from the theory in physics was closed, the economists were obliged to assume that market systems are closed. And because the sum of energy in these equations was conserved, they were also obliged to assume that the sum of utility in market systems is conserved. None of the economists seemed to realize that that utility, or economic satisfaction and well-being, is not comparable to energy as that term was defined in mid–nineteenth century physics. They also failed to realize that market systems are not closed and that the sum of income and utility in any real economic system is not conserved. Nevertheless, these assumptions are foundational to neoclassical economic theory in its current form—constrained maximization in general equilibrium theory.

In the mathematical formalism that resulted from these substitutions, the economic actor is presumed to operate within a field of force identified, in both figurative and literal terms, with energy. Because energy/utility in this mathematical formalism is conserved, the economists realized that production and consumption of goods and commodities must be viewed as physically neutral processes that do not alter the sum of utility. In an effort to justify this idea, they appealed to a very strange interpretation of a law in classical physics which is no longer viewed as valid—the law of conservation of matter, or, the idea that matter cannot be created or destroyed. If matter, claimed the economists, is immutable, then the production of goods and commodities cannot alter or change the basic stuff out of which they are made. And this became the basis for the claim that the immutable stuff out of which goods or commodities are made cannot be changed by consumption and that any value associated with consumption must reside in the minds of economic actors. Strangely enough, this was the origin of two assumptions in neoclassical economic theory that lie buried beneath the maze of mathematical formalism now used by mainstream economists: (1) economic actors interact within a field of force (utility) in which the natural laws of economics legislate over their economic decisions and determine the value of goods, commodities, and services and (2) the value of these goods, commodities, and services circulates in this field as capital in a closed loop from production to consumption in a domain of reality that is separate and distinct from the external environment.

Numerous well-known mathematicians and physicists told the economists that the economic constructs were utterly different from the physical variables and that it was not possible to assume that the constructs were in any sense comparable to the variables. But the economists refused or, more probably, failed to comprehend, how devastating this criticism was and proceeded to claim that their mathematical theories were scientific and very similar to those in physics. In what is surely one of the most bizarre chapters in intellectual history, the origins of neoclassical economic theory in mid–nineteenth physics were forgotten, subsequent generations of mainstream economists disguised the unscientific assumptions about the dynamics of market systems under an increasingly more elaborate maze of mathematical formalism, and the totally unsubstantiated claim that this discipline was scientific was almost universally accepted.[4]

THE CREATORS OF NEOCLASSICAL ECONOMICS

William Stanley Jevons studied chemistry and mathematics in London and attended some of Michael Faraday's lectures at the Royal Institution. In these lectures, Faraday demonstrated that magnetic forces did not obey the Newtonian force rule and claimed that other forces must be present. Jevons was familiar also with the work of William Thomson (Lord Kelvin) and James Prescott Joule on the interconvertibility of heat and mechanical energy that laid the foundations for the law of conservation of energy. But Jevons was not a skilled mathematician, and his understanding of scientific matters was crude at best and completely distorted at worst.

In order to appreciate how the creators of neoclassical economic theory attempted to justify the claim that there was scientific justification for substituting utility for energy in the equations of the theory in physics, consider the following passage from Jevons's major work, *The Principles of Science*: "Life seems to be nothing but a special form of energy which is manifested in heat and electricity and mechanical force. The time may come, it almost seems, when the tender mechanism of the brain will be traced out, and every thought reduced to the expenditure of a determinate weight of nitrogen and phosphorous. No apparent limit exists to the success of the scientific method in weighing and measuring, and reducing beneath the sway of law, the phenomena of matter and mind.... Must not the same inexorable reign of law which is apparent in the motions of brute matter be extended to the human heart?"[5]

Mind, according to Jevons, is a manifestation of energy; the physical substrate of mind can be reduced to a measurable quantity, such as the "weight of nitrogen and phosphorous"; and the "phenomena" of mind are potentially explainable in terms of collections of particles subject to the "inexorable reign" of deterministic physical laws. If one actually believes, as Jevons apparently did, that this is the case, it does not require a great leap of faith to arrive at the conclusion that the collection of particles in the mind of an economic actor is subject to the deterministic natural laws that allegedly operate within the protean field of utility–energy described in the equations of his mathematical theory.

In the following passage from *Theory of Political Economy*, Jevons defends the claim that his theory is scientific: "It is clear that Economics, if it is to be a science at all, must be a mathematical science. There exists much prejudice against attempts to introduce the methods and language of mathematics into any branch of the moral sciences.... My theory of Economics is purely mathematical in character.... The theory consists of applying differential calculus to the familiar notions of wealth, utility, value, demand, supply, capital, interest, labour, and all the other quantitative notions belonging to the daily operations of industry.... To me it seems that our science must be mathematical, simply because it deals in quantities. Whenever things treated are capable of being greater or less, there the laws and relations must be mathematical in nature."[6]

Jevons claimed that his theory is scientific because it describes the behavior of economic actors in terms of well-defined quantities with the use of the differential calculus. But in physics, the differential calculus describes the movement of masses of objects in vector space in terms of continuous functions that result in infinitely small differentials in accordance with the classical laws of motion. Jevons, however, seems to have been quite oblivious to the fact that decisions made by economic actors cannot be described in this fashion. And one does not have to be a trained logician to appreciate the absurdity of his circular argument—the theory must be scientific because it is mathematical and the theory must be mathematical because it is scientific.

Léon Walras was encouraged by his father to study engineering and enrolled in the Ecole des Mines in 1845. Dissatisfied with the study of engineering, Walras read philosophy, history, literary criticism, and political economy. During this period, he also read a popular account of the philosophy of Kant and embraced a confused monism that was a synthesis of materialism and spiritualism.[7] Despite his lack of training in either mathematics or physics, Walras viewed Newtonian physics as the unequaled model of scientific knowledge, and his grand ambition was to use this model to create "the science of economic forces, analogous to the science of astronomical forces."[8]

After substituting utility for energy in the equations of the theory in physics, Walras, like Jevons, was obliged to use an additive utility function in which the utility of a good is solely the function of the quantity of the good consumed. The economists included this function because, like energy in the original equations, utility in the differential calculus borrowed from the theory becomes progressively smaller over time. In an effort to explain why this is the case, the economists claimed that the utility (economic satisfaction and well-being) experienced by an economic actor in the consumption of increasingly larger amounts of a particular good gradually diminishes. What is important to realize here is that the economists included the additive utility function not because it has anything to do with the pleasure derived in the consumption of a good by an economic actor. They were obliged to include it to justify the claim that the mathematical formalism borrowed from the theory in physics discloses the lawful dynamics of market systems.

The additive utility function allowed the utility of a bundle of goods to be expressed as the sum of single utility functions that represents the pleasure derived by the economic actor in the consumption of increasing amounts of goods in the bundle. Walras's *rarete*, which is the equivalent to Jevons's marginal utility, refers to the last infinitesimal increment of utility (pleasure) an economic actor derives from the consumption of a particular bundle of goods. The classic example of how this allegedly lawful dynamic of market systems operates is that the first piece of bread consumed by a hungry man has the most utility; the last piece, the least.

In *Elements of Pure Economics*, Walras asserts that the "theory of industry is called applied science or art; the theory of institutions moral science or ethics."[9] He then claims that the domain of reality in which industry exists is separate and distinct from all other domains, including that of government. And this becomes the basis for the conclusion that the only natural phenomenon in the domain of reality where industry exists is the "single relation between two things represented by the value or price of a good." Walras then claims that "any value in exchange, once established, partakes of the character of a natural phenomenon, natural in origins, natural in its manifestations and natural in essence."[10]

Assuming that "natural" means "from or pertaining to nature," on what basis does Walras conclude that a value established in an exchange can be viewed as natural in its origins, manifestations and essence? Markets are human inventions that have taken a wide range of different forms, the value of any commodity is normally a function of a staggering array of variables, and prices paid are invariably tied to individual tastes and preferences. The answer is that Walras assumed that prices paid by economic actors are determined by forces associated with the natural laws of economics, and this is the basis for his claim that prices are "natural" in origins, manifestations, and essence.

When contemporary mainstream economists use the term *natural*, as in natural rates of unemployment, they rarely comment on its meaning. Their use of the word implies, however, that forces associated with the natural laws of economics are natural and good because they contribute to the growth and expansion of market systems. And this serves to reinforce the view that outside intervention by government or any other agency is unnatural and bad because it interferes with the forces associated with the natural laws.

In spite of the fact that market systems as Walras conceives them are very rigid and highly mechanistic, he does not claim that human will has no influence on prices. But he does say that the forces that regulate comparative prices are comparable to the law of gravity. Just as the force of human will can resist the force of gravity, it can also, according to Walras, resist the forces that regulate competitive prices. But he qualifies this claim by stating that these forces govern the interaction of economic actors in much the same way that gravity governs the interactions of masses of objects. And since the force of gravity in classical physics inexorably moves physical systems toward equilibrium, Walras concludes that economic forces inexorably move competitive prices toward equilibrium.[11]

Francis Ysidro Edgeworth and Vilfredo Pareto also claimed, like Jevons and Walras, that they had transformed the study of economics into a rigorously mathematical scientific discipline. The enthusiasm with which Edgeworth preached this gospel is apparent in the following passage: "The application of mathematics to the world of the soul is countenanced by the hypothesis (agreeable to the general hypothesis that every psychical phenomena is the concomitant, and in some the sense the other side of a physical phenomena), the particular hypothesis, adopted in these pages, that Pleasure is the concomitant of Energy. Energy may be regarded as the central idea of Mathematical Physics: maximum energy the object of the principle investigations in that science.... As the movements of each particle, constrained or loose, in a material cosmos are continually subjugated to one maximum sub-total of accumulated energy, so the movements of each soul whether selfishly isolated or linked sympathetically, may continually be realizing the maximum of pleasure."[12]

Pareto's position, though more pugnacious, is essentially the same as that of Edgeworth: "Strange disputes about predestination, about the efficacy of grace, etc., and in our own day incoherent ramblings on solidarity show that men have not freed themselves from these daydreams which have been gotten rid of in the physical sciences, but which still burden the social sciences.... Thanks to the use of mathematics, this entire theory, as we develop it in the Appendix, rests on no more than a fact of experience, that is, on the determination of the quantities of goods which constitute combinations between which the individual is indifferent. The theory of economic science thus acquires the rigor of rational mechanics."[13]

The wholesale abuse by these economists of a theory in mid–nineteenth physics explains why neoclassical economic theory and the mathematical theories now used by mainstream economists are predicated on the following assumptions:

- The market is a closed circular flow of capital between production and consumption with no inlets or outlets.
- Market systems exist in a domain separate and distinct from the external environment.
- The natural laws of economics, if left alone, will ensure that market systems will perpetually grow and expand.
- The unimpeded operations of the natural laws of economics will result in the perpetual expansion of national economies and the global market system.
- Environmental problems result from market failures or incomplete markets.
- The natural laws of economics can resolve environmental problems via price mechanisms and more efficient technologies and production processes.
- Inputs of raw materials into the closed market system from the external environment are "free" unless or until costs associated with their use are internalized within the system.
- The resources of nature are largely inexhaustible, and those that are not can be replaced by other resources or by technologies that minimize the use of the exhaustible resources or rely on other resources.
- The costs of damage to the external environment by economic activities must be treated as costs that lie outside the closed market system or as costs that are not included in the pricing mechanisms that operate within these systems.

- These costs can be internalized in the closed market system with the use of shadow pricing and the establishment of property rights for environmental resources and amenities.
- There are no biophysical limits to the growth of market systems.

Obviously, these assumptions make no sense at all in scientific or ecological terms. Markets are open systems that exist in embedded and interactive relationship to the global environment, and there is a very definite relationship between economic activities and the state of this environment. Natural resources are clearly exhaustible, and our overreliance on some of these resources, particularly fossil fuels, could soon result in irreversible large-scale changes in the climate system. The natural environment is not separate from economic processes, and wastes and pollutants from these processes are already at levels that are massively disrupting virtually all environmental subsystems. Last but not least, the limits to the growth of the global economy in biophysical terms are real and inescapable, and the assumption that market systems can perpetually expand and consume more scarce and nonrenewable natural resources is utterly false.[14]

A GREEN THUMB ON THE INVISIBLE HAND

When mainstream economists are confronted with the charge that there is no basis in neoclassical economic theory for implementing scientifically viable economic solutions for environmental problems, they typically deny that this is the case by appealing to the work done by environmental economists. This orthodox approach to resolving environmental problems is taught in universities and practiced in government agencies and development banks, and the solutions are almost invariably embedded in the mathematical formalism of general equilibrium theory.

Because the practitioners of neoclassical economic theory assume that the gross national product in functional market economies must expand by at least three percent per annum, environmental economists presume that the health of these economies is sensitively dependent on the consumption of increasingly larger amounts of environmental resources. And because the theory is predicated on the assumption that market systems are closed and exist in a domain of reality separate and distinct from natural resources in the so-called external environment, environmental economists presume that these resources are not subject to the pricing mechanisms that operate within these systems.

When environmental economists calculate the environmental costs of economic activities, they assume that the relative price of "each bundle" of an environmental good, service, or amenity reveals the "real marginal values" of the consumer. Note what the writers of a standard textbook on environmental economics have to say about the dynamics of this process: "The power of a perfectly functioning market rests in its decentralized process of decision making and exchange; no omnipotent planner is needed to allocate resources. Rather, prices ration resources to those that value them the most and, in doing so, individuals are swept along by Adam Smith's invisible hand to achieve what is best for society as a collective. Optimal private decisions based on mutually advantageous exchange lead to optimal social outcomes."[15]

In environmental economics, the belief that optimal private decisions "based on mutually advantageous exchange" lead to "optimal social outcomes" for the state of the environment is a primary article of faith. But according to these economists, this will not occur unless the following conditions apply: the market system in which economic actors make optimal private decisions must operate more or less perfectly, and the prices, or values, of environmental goods and services must be represented as a function of those decisions. But if these conditions are met, environmental economists presume that the lawful dynamics of market systems associated with the invisible hand will resolve environmental problems when the "prices are right."

According to these economists, the right price is the price that economic actors have paid, or are willing to pay, to realize some marginal benefits of consuming environmental goods and services. Because the right price in neoclassical economic theory is determined by forces that operate within closed market systems, environmental economists view natural resources that cannot be valued in these terms as "environmental externalities." And they define externalities as applying to situations in which the production or consumption of one economic actor affects another who did not pay for the good produced or consumed.

According to environmental economists, externalities are either negative or positive. Pollution is often cited as an example of the former and preservation of biological diversity as an example of the latter. When these economists use the phrase "environmental externalities," they are referring to environmental goods and services that are "external" to market systems in the sense that they are presumed to exist outside of the domain in which the dynamics of these systems allegedly govern decisions made by economic actors and determine the right price.

Environmental economists often use cost-benefit analyzes different from those used by other mainstream economists to place a value on environmental externalities. The problem that these accounting procedures are intended to resolve is that the only "real" marginal values the environmental economists can confer on the environment are determined by forces associated with the natural laws of economics that operate within closed market systems. Given that the vast majority of the damage done to the natural environment by economic activities cannot be valued in these terms, environmental economists have developed indirect methods designed to estimate the "use-value" of these natural resources.[16]

For example, environmental economists use the travel cost method to assess the use value of nonmarket resources, such as national parks and public forests, and to determine the "willingness to pay function" of those who consume these resources. In this method, a statistical relationship between observed visits to nonmarket resources of natural beauty and the costs of visiting those resources is derived and used as a surrogate demand curve from which the consumer's surplus per visit–day can be measured. While the travel cost method of evaluation may seem rather esoteric and quite strange, it has been widely used to assess the costs and benefits of proposals to create or preserve publicly owned recreational areas in the United States and Britain.[17]

Environmental economists use "contingent valuation surveys" to assess the use value of the nonmarket resources of recreation, scenic beauty, air quality, water quality, species preservation, and bequests to future generations. The word "contingent" is meant to highlight the fact that the values disclosed with the use of these surveys are contingent on the artificial or simulated market conditions described in the surveys. The intent of these surveys is to determine the amount that economic actors might be willing to pay to preserve natural environments (preservation or existence values), maintain the option of using natural resources (option values), and bequeath natural resources to future generations (bequest values).[18] This is normally accomplished by asking the respondents to indicate the maximal amount they are willing to pay for an increase in the quality of an environmental resource and the minimal amount they are willing to accept as compensation to forgo this increase.

For the sake of argument, let us assume that contingent valuation studies provide a basis for realistically assessing the optimal social outcomes of environmental policy decisions. Are we then to believe, as one contingent valuation showed, that reduction in chemical contaminants in drinking water was not important in economic terms because the value of a statistical life associated with a reduction in risk of death in thirty years was only $181,000?[19] Is $26 a measure of the real marginal costs of pollution because this is the average price that a household is willing to pay annually for a ten percent improvement of visibility in eastern U.S. cities?[20] Is the value of whopping cranes the $22 per year average that one set of households was willing to pay to preserve this species[21] and that of the bald eagle the $11 per year average that another set of households would spend to preserve this apparently less valuable species?[22]

The obvious question here is how could environmental economists possibly assume that the amount of money that people who have a wide range of educational levels and average incomes might be willing to pay to preserve an environmental resource or resolve an environmental problem can serve as the basis for implementing public policies that have optimal social outcomes? They do so by appealing covertly or overtly to two unscientific assumptions embedded in the mathematical formalism used by these economists. The first is that market systems exist in a field of amorphous and protean utility/energy in a domain of reality separate and distinct from the external environment. And the second is that forces in this field govern decisions made by economic actors and determine the right prices.

The absurdity of the assumption that market systems exist in a domain of reality separate and distinct from the external environment is painfully apparent in an article written by a well-known environmental economist on the potential economic impacts of global warming. After concluding that "climate change is likely to have different impacts on different sectors in different countries," the author says the following about the U.S. economy: "In reality, most of the U.S. economy has little interaction with climate. For example, cardiovascular surgery and parallel computing are undertaken in carefully controlled environments and are unlikely to be directly affected by climate change. More generally, underground mining, most services, communications, and manufacturing are sectors likely to be largely unaffected by climate change—sectors that comprise about 85 percent of GNP."[23]

Obviously, there is no basis for assuming that sectors of an economy can be isolated from the impacts of global warming because they have little or no "interaction" with climate. If average earth temperature increases by three or four degrees Centigrade, which now seems very likely, global warming would have disastrous impacts on environmental systems in virtually all regions or territories on the planet, including those in United States. And since these impacts would massively disrupt production, distribution, and transportation systems in all sectors of the U.S. economy, the claim that sectors that now "comprise about 85 percent of GNP" will be "largely unaffected by climate change" is patently absurd.

A fair number of economists over the past two decades, including luminaries like Arrow and Hand, have expressed doubt about the efficacy of assumptions about the dynamics of market systems in neoclassical economic theory. And some theoretical economists, such as the game theorists and the practitioners of behavioral economics, have challenged the validity of these assumptions. But the vast majority of economists in business and government are not interested in the most advanced theoretical work in their discipline.

Legions of these economists are engaged on a daily basis in developing analyses and making predictions that guide the decision making of political leaders and economic planners. Most of these individuals are aware that the resulting economic activities could have destructive environmental impacts and try to minimize these impacts as long as profit margins can be maintained. But these good intentions are hugely ineffectual because the mathematical theories used by these economists disallow the prospect of realistically assessing the environmental costs of economic activities and internalizing these costs in pricing systems.

Mainstream economists do not view themselves as priests in the temple where the true believers in the absentee deistic god with the invisible hand have given them the power to legislate over the future of life on planet earth. But this is a fairly accurate description of the role played by these economists in the process of developing and implementing economic solutions for environmental problems. And the status of the priests in the temple of neoclassical economic theory has been greatly enhanced by the widespread belief that this theory is scientific.

The next chapter will demonstrate that this belief not only massively contributed to the meltdown of the global financial markets in the fall of 2008. It also served to legitimate a program of economic globalization known as the market consensus, which is predicated on the assumption that the lawful dynamics of market systems will result in a new global order. In this vision of the human future, all national economies will be free market systems and the governments of all nation-states will operate in accordance with the principles of democratic capitalism. One large problem here is that the market consensus sanctions and promotes the growth and expansion of the fossil fuel–based global market system. The other is that if this system continues to grow and expand over the next two decades, global warming will trigger irreversible large-scale changes in the climate system, and the terms of human survival will be very harsh indeed.

CHAPTER 8

The Old Story: Economic Globalization, the Market Consensus, and the New State Religion

> I am beginning to think that for all the religions of the world, however they may differ from one another, the religion of The Market has become the most formidable rival, the more so because it is rarely recognized as a religion.
>
> Harvey Cox

The New York Times editorial page attributed the lack of regulation that resulted in the meltdown of the financial markets in 2008 to the "Bush administration's magical belief that the market, with its invisible hand, works best when it is left alone to self regulate and self correct."[1] But what the editorial failed to mention is that the Bush administration's $700 billion economic stimulus plan and the Obama administration's $789 billion American Recovery and Reinvestment Act were both predicated on this magical belief. The fundamental assumption in these plans was that the meltdown occurred because the self-correcting and self-regulating dynamics associated with the invisible hand ceased to function properly. And the intent of the plans was to create market conditions in which these dynamics could begin to function properly with a massive infusion of capital generated by deficit spending.

This meltdown began after the collapse of the markets for derivative contracts that allow buyers to hedge against economic gains or losses. In the parlance of mainstream economists, a derivative is an agreement between two parties that the value of something is determined by the price movement of something else, and hedging allows a buyer or seller to protect assets or incomes against future rises in prices. In derivatives markets, debt is used to generate surplus capital, and this surplus is used to borrow increasingly larger sums of money in a process economists call financial leveraging.

Traditional derivative trading was in commodity-related futures contracts, and the amount of debt that could be used as financial leverage was highly regulated. In these markets, buyers could hedge against unpredictable changes in the prices of real assets, such as wheat or cotton, and each commodity was traded separately. But this situation changed dramatically after December 2000, when the U.S. Congress banned the regulation of derivatives by passing the Commodity Futures Modernization Act.

The rationale for passing this bill, which was largely written by representatives of the investment banks that would later make enormous profits in derivatives trading, appealed to two assumptions in neoclassical economic theory. The first was that the dynamics of closed market systems in derivatives trading are self-correcting and do not need or require regulation by the exogenous agency of government. And the second was that government regulation of this trading interferes with these dynamics and impairs the growth and expansion of the derivatives markets.

Deregulation resulted in the rapid creation of new forms of derivatives contracts, and one of the innovations that had disastrous consequences was collateralized debt obligations (CDOs). In this derivatives market, subprime mortgages were sold by lenders to the investment banks and the banks bundled the mortgages with other loans, including car loans, student loans, and credit card debt. The banks then paid the big three rating agencies, Standard & Poor's (S&P), Moody's, and Fitch, to evaluate the CDOs, and most were given the highest possible rating. The investors who bought shares in the CDOs assumed that this rating indicated that there was a high probability that the debts would be paid and that they would reap enormous profits.

Meanwhile, the largest insurance company in the world, American International Group (AIG), began to sell credit default swaps, which allegedly would function as a kind of insurance policy protecting investors from potential losses. Investors in the credit default swaps paid AIG a quarterly premium, and the insurance company promised to compensate them for potential losses. Trading in CDOs and credit default swaps was highly leveraged, in many instances forty to one. And since the investment banks and AIG did not have sufficient capital in reserve to cover the potential losses, even a small decrease in the asset base could make them insolvent. Not long after these decreases began to occur, one of the investment banks, Lehman Brothers, declared bankruptcy and AIG collapsed. It soon became clear that a meltdown in the global financial markets that could result in a worldwide depression more severe and prolonged than in the 1930s was under way.

Almost two years after the meltdown, Congress passed the Dodd Frank bill, which created new agencies that have oversight over derivatives trading and the authority to impose restrictions on risky investments. But the true believers in the magical machinations of the invisible hand in the Congress refused to properly fund these agencies and have gutted most of the major provisions of the bill. The unfortunate result is that traders in the derivatives markets are doing business as usual, and these markets are still dominated by a small number of investment banks that are too big to be to be allowed to fail.

The global derivatives market is now valued at more than a thousand trillion dollars, and none of these resources is being directly invested in enterprises that involve the production, distribution, and exchange of goods, commodities, and services. The unpredictable events that now can be hedged against include changes in the price of tea in China, the amount of revenue that might be generated by a film, fluctuations in the exchange rate, and the potential passage by Congress of a piece of legislation that could have economic impacts.

One of these derivative exchanges, the Policy Analysis Market (PAM), was created by the Pentagon in an apparent attempt to enlist the magical machinations of the invisible hand in forecasting political events in the Middle East. The menu of unpredictable events that buyers could hedge against on PAM included terrorist attacks, assassinations of political leaders, and cross-border conflicts that could escalate into full-scale wars. But after some news commentators pointed out that this government-sponsored exchange would allow profits to be made by placing bets on when the next terrorist attack would occur, PAM was shut down one day after it was launched.[2]

There is also a market for derivatives trading that allows buyers to hedge against unpredictable weather events linked to global warming. In one form of these contracts, two firms that face diverse economic risks from extreme weather conditions can enter an agreement stipulating that payments will be made to one or the other party depending on whether the temperature moves above or below a specific threshold. In another form known as weather-based call and put options, the buyer receives a payment if the average temperature over a given period of time falls above or below a specified threshold.

What the buyers of these contracts have apparently failed to realize is that the cost benefit analyses of hedging against the uncertainties of extreme weather events are predicated on unscientific assumptions about the dynamics of market systems in neoclassical economic theory. One of these now-familiar assumptions is that these systems exist in a domain of reality separate and distinct from the global environmental and can continue to grow and expand regardless of conditions in this environment. Another is that the environmental impacts of global warming on market systems will have adverse economic impacts on some economies and not on others.

One large problem here that the economists who did these costs benefit analyses apparently did not see or chose to ignore is that market systems in the real world are embedded in and interactive with the global climate system. The other is that self-reinforcing positive feedback loops between irreversible large-scale changes in this system and environmental subsystems will have disastrous impacts on all economies. And since we have now entered a period where extreme weather events will increase dramatically, it is reasonable to conclude that the costs of compensating the buyers of the weather-related derivatives contracts will soon escalate to the point where these markets will collapse.

Adam Smith assumed that decisions made by economic actors in the pursuit of self-interest could benefit all members of society because he believed that the natural laws of economics were created by a benevolent, albeit absentee, deistic god. But derivative contracts do not provide any concrete benefits to society, and this is painfully obvious in the weather related contracts. The intent of these contracts is to allow buyers to place bets on the hugely destructive environmental impacts of global warming in a global casino where the odds makers claim that they can reap enormous profits. And the players in this casino apparently feel that there is nothing wrong with beating the odds and making these profits in spite of the fact that the lives of hundreds of millions of people will be threatened by the extreme weather events.

Many economists have warned that business as usual in derivatives trading will result in another meltdown of the financial markets. But what these economists apparently do not grasp is that this crisis would occur during a period in which the scope and scale of extreme weather events is increasingly dramatically. And since these events will massively disrupt the production and distribution of goods, commodities, and services in the global market system, the next crisis in the financial markets could easily result in the collapse of this system.

THE GOD WITH THE INVISIBLE HAND AND THE MARKET CONSENSUS

Belief in the magical machinations of the invisible hand also resulted in a program of economic globalization that could soon make it impossible to prevent the most disastrous consequences of global warming. This program is known as the market consensus, and it was articulated and implemented during the first administration of President Ronald Reagan. The market consensus is predicated on the assumption that the magical machinations of the invisible hand will necessarily result in a new global order in which all economies are free market systems and the governments of all nation-states operate in accordance with the principles of democratic capitalism.

The back story about the market consensus involves a heated and prolonged debate about the dynamics of market systems in the work of two mainstream economists—liberal economist John Maynard Keynes and conservative economist Milton Friedman. Keynes became internationally known in the aftermath of World War II, when there was no private sector capable of mobilizing the investment, capital goods, and skills required to rebuild the national economies devastated by this conflict. Only governments seemed capable of marshaling the resources needed to deal with these large problems, and the economic model used in most industrial nations in the West and in large parts of the developing world was based on Keynes's conception of reformed and managed national economies.

In these "mixed economies," state ownership, industrial policy, and fiscal management were used in various combinations to protect capitalism from its own excesses and to save capitalism from the lure of socialism. Until the late 1970s, Keynesian "new economics," with its emphasis on managing the overall economy with the fiscal tools of taxation and spending, appeared, for the most part, to have fulfilled its promise of sustained economic growth and full employment. Many economists challenged Keynes's vision, but the most fervent anti-Keynesian, whose name would become a household word during the Reagan era, was Milton Friedman.

After completing graduate work at the University of Chicago, Friedman joined the faculty there in 1946. During the 1950s he became the most vocal opponent of the economic theory of Keynes and the then very influential Keynesian economists at Harvard University. Friedman and the other economists in the Chicago school believed that a small number of mathematical theorems could predict allocations of resources and the pricing of these resources. And since this formalism was premised on "fundamental" assumptions about the dynamics of market systems in neoclassical economic theory, the economists at the Chicago school became known as market fundamentalists.

In response to the claim by Keynesian economists that the mathematical theories used by the Chicago economists were dogmatic, rigid, and premised on a simple-minded understanding of the complex dynamics of market systems, Friedman launched a counteroffensive. In *Capitalism and Freedom*, published in 1962, he argued that market systems cannot function properly in the absence of economic freedom and that this freedom cannot exist in the absence of political liberty.[3] This marked the beginning of Friedman's celebrity status among conservatives, and that status was considerably enhanced when he served as the principal economic advisor to Republican presidential candidate Barry Goldwater in 1964. After Goldwater lost this election, Freidman popularized his understanding of the dynamics of market systems in the mass-market best-seller *Free to Chose* and a related series of programs on public television.[4]

A few years later, Friedman retired from teaching, joined the Hoover Institution at Stanford, and established direct contact with Ronald Reagan and his advisors. After Reagan was elected president in 1980, the true believers in market fundamentalism, the neoconservatives, became a powerful force in American politics. And during the eight years that Reagan was in office, the neoconservatives appealed to market fundamentalism to legitimate a program for economic globalization that became known as the market consensus.

The demise of the Soviet Union in 1989 convinced many political analysts, news commentators, and politicians in the United States that the process of economic globalization was realizing the vision of the human future in the market consensus. These politically powerful individuals appealed to this vision to explain why the United States won the Cold War and was now the last remaining superpower. And the widespread acceptance of this view resulted in the creation of a new state religion that featured as one of its basic doctrines a revised conception of the manifest destiny of the United States.

In this new conception of manifest destiny, the United States is destined to become the unchallenged and undisputed superpower in the new global order envisioned in the market consensus. Traditional conceptions of the manifest destiny of the United States were premised on the belief that this destiny is willed by God and part of a sacredly ordained providential plan. But in the new conception of manifest destiny, the natural laws of economics are the laws of God and the new global order that will allegedly result from the operations of these laws is part of a sacredly ordained providential plan.

IN THE MARKET WE TRUST

Numerous writers of articles and books on economic globalization embraced the new state religion and actively promoted the revised conception of manifest destiny of the United States. *New York Times* columnist Thomas Friedman was among the first to claim that the new global order envisioned in the market consensus would be realized through the process of economic globalization. In one of his columns, Friedman argued that although people once thought that human affairs could be ordered in the absence of free markets, this was no longer the case in the new era of globalization: "I don't think there will be any an alternative ideology this time around. There are none."[5]

The thesis in the first of Friedman's best-selling books, *The Lexus and the Olive Tree*, is that the forces associated with the dynamics of market systems will transform the governments of all nation-states into functional democracies. These forces are responsible, claimed Friedman, for the "democratization of technology," the "democratization of finance," and the "democratization of information."[6] At several points in this hugely popular book, Friedman clearly implies that the natural laws of economics are created by God and asks in the last chapter how a "visionary geo-architect," or God, would design an ideal nation-state. The answer, says Friedman, is that this nation-state would be the "most flexible market in the world."[7]

Francis Fukuyama, in another best-seller, *The End of History*, does not claim that the process of economic globalization is proceeding in accordance with a sacredly ordained plan. But he does argue that this process is inexorably moving the global community toward the "end point of mankind's ideological development and the final form of human government."[8] According to Fukuyama, the forces of democratic capitalism have already "conquered rival ideologies like hereditary monarchy, fascism, and most recently communism." And he also claims that when this process is complete, history will end in the sense that a "single coherent, evolution process" will result in a new global order in which all governments are "liberal democracies" and all economies are linked to the global market system.[9]

In other best-selling books promoting the vision of the human future in the market consensus, the conflation of the natural laws of economics with the laws of God is quite obvious and virtually impossible to ignore. For example, George Gilder, the author of numerous popular books on business, states in *The Sprit of Enterprise* that, "It is the entrepreneur who knows the rules of the world and the laws of God."[10] In *Wealth and Poverty*, Gilder declares that virtually all societal problems in the United States can be resolved by the unfettered operation of the natural laws of economics.[11] In Gilder's view, these laws were created by God, and the failure to recognize that this is the case is the principal cause of virtually all economic and social problems in the United States and, by implication, in other countries as well.

Kevin Kelly, in yet another best seller, *Out of Control: The New Biology of Machines, Social Systems and the Economic World*, says that technological innovation is proceeding in accordance with God's providential plan. And he refers to his own list of the dynamics of market systems as the "Nine Laws of God."[12] In an article entitled "The Market 'R' Us," the economist Robert Samuelson claims that the dynamics of the market system always act in the best national interests of the United States, whether we know it or not, and against the interests of countries that do not embrace this system.[13] Robert Bartley, editor of the *Wall Street Journal*, championed a similar view. According to Bartley, the "world is not ruled by politicians but by markets," and national governments "will evolve toward something like state governments today. Each will have its own industrial development program to show why it has the best business and investment climate."[14]

Over the past two decades, increasingly larger numbers of Americans have embraced this new state religion and the associated doctrine of the manifest destiny of the United States. This is apparent in the rhetoric used by members of the Tea Party, by the neoconservatives in the Congress, by right-wing talk show hosts on radio and television, and by the climate change contrarians. It is apparent also in the fact that many, if not most, Americans assume that the unscientific truths about the dynamics of market systems in the neoclassical economic theory have more validity and greater authority than the truths of science.

THE RELIGION OF THE MARKET

Theologian Harvey Cox makes a convincing case that there is a striking resemblance between the "lexicon of *The Wall Street Journal* and the business sections of *Time* and *Newsweek*" and "Genesis, the Epistle to the Romans, and Saint Augustine's City of God." He claims that what lies beneath the commentary in these publications on "market reforms, monetary policy, and the convolutions of the Dow" is "a grand narrative about the inner meaning of human history, why things have gone wrong, and how to put them right." According to Cox, this grand narrative is predicated on a "comprehensive business theology" that features "sacraments to convey salvific power to the lost, a calendar of entrepreneurial saints, and what theologians call an eschatology—a teaching about the end of history."

Cox refers to the deity at the "apex of this theology system" as the "The Market" and notes that the capital letters are meant to "signify both the mystery that enshrouds it and the reverence it inspires in business folk." Like God in the Judeo Christian tradition, The Market, says Cox, is "omnipotent (possessing all power), omniscient (having all knowledge) and omnipresent (existing everywhere)." Like Yahweh in the Old Testament, The Market is the "Supreme Deity, the only true God, whose reign must be universally accepted and who allows no rivals." And like the Creator in the book of Genesis, The Market has the power to "define what is real, to make something out of nothing and nothing out of something," and there is "no conceivable limit to its ability to convert creation into commodities."

Toward the close of this argument, Cox makes the following comment: "I am beginning to think that for all the religions of the world, however they may differ from one another, the religion of The Market has become the most formidable rival, the more so because it is rarely recognized as a religion." One of the major concerns here, says Cox, is that while nature in traditional religions is sacred, there is nothing sacred about nature in the religion of The Market. In this religion, "human beings, more particularly those with money, own anything they buy and—within certain limits—can dispose of anything they chose." Cox concludes this argument with the comment that while "all traditional religions teach that all human beings are finite and that there are limits to any earthly enterprise," the "First Commandant" of the religion of The Market is, "There is never enough."[15]

David Loy also makes a convincing case that mainstream economic theory is "a theology pretending to be a science" and claims that "the Market is becoming the first truly world religion, binding all corners of the globe into a worldview and set of values whose religious role we overlook only because we insist on seeing them as secular."[16] According to Loy, the "ecological catastrophe is not only awakening us to the fact that we need a deeper source of values and meanings than market capitalism can provide, but also to the realization that contemporary religion in not meeting this need either."[17] Another must read for those interested in learning more about the prophets, the priesthood, the sacred texts, the core doctrines, and the plan of salvation in the religion of the market is Robert Nelson's book *Economics as Religion*.[18]

TOWARD A NEW THEORY OF ECONOMICS

Fortunately, there are economists in a discipline known as ecological economics who are very much aware that neoclassical economic theory is predicated on unscientific assumptions about the dynamics of market systems. The mistake, if one can call it that, made by the ecological economists is the presumption that mainstream economists would be willing to revise these assumptions in an effort to implement scientifically viable solutions for environmental problems. Given the enormous extent to which these assumptions have contributed to the environmental crisis and are frustrating its resolution, there is obviously nothing unreasonable about this presumption. But the fact that there has been virtually no dialogue between ecological economists and mainstream economists, including environmental economists, clearly indicates that the former is saying something that the latter simply does not wish to hear.

Perhaps the best way to explain why the ecological economists have been systematically dismissed or ignored by mainstream economists is to briefly consider the work of a seminal figure in this discipline—Herman Daly. According to Daly, neoclassical economic theory is a program for ecological disaster because it fails to take into account the first and second laws of thermodynamics. The first law states that energy may be transformed, or changed from one state to another, but cannot be created or destroyed. And the second states that low-entropy matter–energy in a closed system is always transformed into high-entropy matter–energy.

Entropy in physics is essentially a measure of disorder in a physical system—the higher the entropy, the greater the disorder. From the perspective of thermodynamics, an economic system converts matter–energy from a state of low entropy to a state of high entropy and matter–energy exists in two forms—available or free and unavailable or bound. For example, the chemical energy in a piece of coal, which is low in entropy, is viewed as free, and the heat energy in waters of the oceans, which is high in entropy, is viewed as bound. Given that the amount of bound matter–energy in a market system must continually increase, the only way to lower entropy in this system is to use more matter–energy. But the price that must be paid for the consumption of more matter energy is an overall increase in entropy.

Virtually every object with economic value has a highly ordered structure, and the matter–energy required to manufacture these objects increases the overall level of entropy. For example, an automobile is vastly more ordered that a lump of iron ore, and the matter–energy required to transform raw materials into an automobile is enormous. Manufacturing processes also produce waste, pollution, and greenhouse gases that massively contribute to the overall level of disorder in the ecosystem. Even recycled products require energy inputs that increase entropy levels and convert free energy to bound energy.

In *Steady State Economics*, Daly explains in great detail why assumptions about the dynamics of market systems in neoclassical economic theory disallow the prospect of accounting for the throughput of low-entropy natural resources. He also makes a convincing case that the emphasis in this theory on money flows, or on the movement of quantities of money over periods of time, serves to perpetuate the fiction that perpetual economic growth is possible and desirable. The solution to these problems, says Daly, is to use constraints associated with the second law of thermodynamics to formulate policies for long-term sustainability, such as taxes on energy and virgin resources. In his view, these policies would increase social awareness of ecological limits and promote the realization that "physical flows of production and consumption must be minimized subject to some desirable population and standard of living."[19]

According to Daly, the three basic goals of an economic system should be efficient allocation, equitable distribution, and sustainable scale. The first two goals are implicit in neoclassical economic theory, and specific public policies have been formulated to realize them. But there is nothing in this theory that can account for scale and no policy instruments that deal with scale. Daly defines scale as the total physical volume of low-entropy raw materials that move through the open subsystem of an economy and back into the finite and nongrowing global environment as high-entropy wastes.[20]

Based on the scientifically valid fact that the scale of the global economy has grown dangerously large relative to the ecosystem, Daly comes to the scientifically valid conclusion that this economy is sustainable only if it does not erode the carrying capacity of the ecosystem. The large problem here, says Daly, is that there is no basis in neoclassical economic for even recognizing, much less dealing with, this scientific truth. One reason why this is the case is the assumption in this theory that environmental resources and sinks are inexhaustible and that decisions about allocation merely move natural resources between alternative uses. Another is that this theory sanctions and promotes public policies that are predicated on the assumption that there are no biophysical limits on the growth and expansion of the fossil fuel–based global market system.

The core problem here is that the value of natural resources in neoclassical economic theory can be determined only by the prices that economic actors have paid or are willing to pay to consume or preserve these resources. And this, says Daly, explains why there is no basis in the pricing mechanisms in this theory for dealing with the fact that distribution and scale "involve relationships with the poor, the future, and other species that are fundamentally social in nature rather than individual." One unfortunate result is that decisions involving these relationships exist on the same plane as the choice between chewing gum and a candy bar. The other is that this absurd assumption is "dominant in economies today" and "part of the retrograde reduction of all ethical choice to the level of personal tastes weighted by income."[21]

For the past three decades, Daly has argued that we must begin to create a "steady state economy." He defines this economy as a global economic system in which there would be "constant stocks of people and artifacts, maintained at some desired, sufficient levels by low rates of maintenance throughput, that is, by the lowest feasible flows of matter and energy from the first stage of production to the last stage of consumption."[22] Some critics of Daly have argued that keeping consumption constant would require harsh government controls. But Daly claims that such an economy could flourish in democratic societies with a commonsense mixture of free markets and market regulations. And the few countries that have attempted to create some semblance of a steady-state economy, most notably Sweden, Denmark, and Germany, have shown this claim to be correct.

Given that Daly privileges market mechanisms, one might suppose that his views would be welcomed by mainstream economics. But this is not the case, for two reasons. First, Daly challenged the validity of the unscientific assumption in the neoclassical economic paradigm that market systems can perpetually grow and expand. And second, his scientifically valid claim that economic systems are open to and interactive with the global environment challenged the assumption in this paradigm that market systems are closed. If mainstream economists admitted that this was the case, they would be obliged to recognize that there is no basis for assuming that the dynamics of market systems legislate over decisions made by economic actors, promote the growth and expansion of these systems, and maintain the systems in relative equilibrium. They would be obliged also to recognize that there is absolutely no basis for assuming that the mathematical theories they used throughout their careers to make a living and enhance their reputations are scientific.

There is now a significant body of research on what will be required to create a steady-state economy. And much of this research has been done by ecological economists Peter Victor and Tim Jackson and by environmental law and public policy expert Gus Speth. These scholars have not only demonstrated that business as usual in the fossil fuel–based global market system is a program for ecological disaster. They have also made a convincing case that greater efficiencies in industrial economies and the widespread implementation of alternate energy resources and green technologies will not in themselves resolve the environmental crisis. When one reads their carefully reasoned and scientifically sound assessments, it becomes clear that this crisis has its roots in the overconsumption of natural resources. And it also becomes clear that this crisis can be resolved only by greatly reducing the amount of material stuff produced and consumed in the global market system.

The work of the environmental economists is quite extensive and replete with scientifically viable economic solutions for environmental problems. And it is these economists who have the background and training required to develop an environmentally responsible economic theory predicated on a well-defined set of scientifically valid assumptions about the relationship between human and environmental systems. In this theory, these assumptions would be embedded in a mathematical formalism that weds baseline measures of sustainability to realistic assessments of the environmental costs of economic activities and provides a basis for including these costs in pricing systems.[23]

This new economic theory, as David Korten said in a 2011 presentation to the Society for Ecological Economics, would be designed to achieve the following objectives: (1) "ecological balance between aggregate human consumption and the regenerative capacity of the biosphere"; (2) "equitable distribution of real wealth to meet the needs of all"; and (3) "a living democracy to secure economic and political accountability to people and community through active citizen participation."[24]

The ecosystem in this theory would be viewed as the source of all life, and preserving and protecting the capacity of this system to sustain the richness and diversity of life would be a sacred and nonnegotiable responsibility. Economic performance would be measured in terms of increases in the health and happiness of people, households, and communities. And the purpose of economic activities would be to ensure the flourishing of life and the well-being of all of the members of the extended human family.

In my view, the greatest barrier to developing and implementing an environmentally responsible economic theory in the United States is that most Americans, along with virtually all of their elected representatives in Congress, are, by varying degrees, true believers in the religion of The Market. This situation is further complicated by the fact that this religion is preached and practiced by executives and board members of international corporations and investment banks, traders in all major stock exchanges, and players in the global casino of the derivatives markets. And as anyone who watched the televised debates during the Republican primary elections of 2012 should know very well, all of the candidates for the nomination of this party to run for president of the United States preach and practice this religion.

What these true believers have apparently failed to realize is that they are unwittingly worshipping at the altar of a false god that is not the God of the Judeo Christian tradition. There is nothing in the sacred texts of this tradition that provides any religious justification whatsoever for believing that this God created the natural laws of economics and declared that the pursuit of self-interest serves the common good. And there is nothing in these texts that suggests that the vision of the human future in the market consensus is willed by God or that the manifest destiny of the United States is part of a sacredly ordained providential plan. It is also quite impossible to find any religious justifications for these beliefs in the sacred texts of other great religions of the world.

The aim in the next and last chapter is to accomplish four objectives. The first is to demonstrate that the most fundamental scientific truths about the relationship between self and world in physical reality are analogous to and compatible with the most profound spiritual truths about this relationship in spiritual reality in the five great religious traditions of the world. The second is to make the case that a new dialogue between the truths of science and religion can serve as the basis for articulating and disseminating an environmental ethos with a profound spiritual dimension. The third is to provide substantive validity to the claim that if sufficient numbers of spiritually aware and environmentally concerned people enter this dialogue, this could result in the fairly rapid emergence of a well-organized and highly effective worldwide movement in religious environmentalism. And the fourth objective is to explain why this movement could be critically important in the effort to resolve the environmental crisis during the relatively short time frame in which this will remain a possibility.

CHAPTER 9

The Dream of This Place with Us: Science, Religion, and the Environmental Crisis

When you come, as you soon must, to the streets of our city,
Mad-eyed from stating the obvious,
Not proclaiming our fall but begging us
In God's name to have self-pity.

Spare us all word of the weapons, their force and range,
The long numbers that rocket the mind;
Our slow, unreckoning hearts will be left behind,
Unable to fear what is too strange.

Nor shall you scarce us with talk of the death of the race,
How should we dream of this place without us?—
The sun mere fire, the leaves untroubled about us,
A stone look on a stone's face?

          Richard Wilbur, "Advice to a Prophet"

The first color photograph of the whole earth taken by an astronaut aboard the Apollo 11 spacecraft in 1969 appeared in posters, popular magazines and films, and a staggering number of advertising campaigns promoting the sale of an enormous variety of consumer goods and services. Like any iconic visual image repeatedly used in a variety of diverse contexts for different purposes, the image of the blue planet shimmering under the flow of its delicate atmosphere against the backdrop of the interstellar dark seemed, over time, to lose much of its original significance and emotive power. But for some of us, myself included, it still serves as sign and symbol of two astonishing facts: life on earth is a self-organizing and self-perpetuating whole that evolved from a single organism to a level of enormous complexity and incredibly beauty; our species emerged from this evolutionary process as fully conscious and self-aware beings in the vast cosmos.

Many scientists are convinced that biological life must have arisen on other planets and that the process of evolution on some of them resulted in the emergence of intelligent life-forms and advanced technological civilizations. The most ardent promoters of this idea are physicists and molecular biologists, and public acceptance of their views resulted in the creation of the Search for Extraterrestrial Intelligence (SETI) project. The intent of those involved in this project, as many Americans learned for the first time after seeing a movie based on the late Carl Sagan's book *Cosmos*, is to intercept intragalactic or intergalactic communications between advanced civilizations on other planets. But what the book and movie failed to mention is that many evolutionary biologists are convinced that the odds that these civilizations exist anywhere else in the vast cosmos are slim to none.

The skeptics argue that even if life exists on many other planets, this does not mean that the process of evolution on any of them would result in life-forms that have the capacity to acquire and use a complex symbol system like the human language system. During billions of years of evolution on this planet, only a small percentage of mammals, the anthropoid apes, emerged via innumerable indeterminate branch points with an intelligence that surpasses that of other mammals. During the twenty-five million years in which these apes have existed, there were probably hundreds of branching points and independently evolving lines, and only one of them became the lineage that evolved into fully modern humans. When we also consider that the evolutionary pathway that allowed our species to acquire and use fully complex language systems was utterly improbable, highly unique, and involved a myriad number of mutations over about two million years, it is not difficult to appreciate why many evolutionary biologists seriously doubt that intelligent life exists on other planets.[1]

The scientific jury is still out on this question, and this will probably remain the case for a very long time. But if the skeptics are correct, the question of whether the fully conscious and self-aware life-forms on this planet will resolve the environmental crisis takes on a much larger significance. If we fail and our species becomes extinct, this would be a tragedy of cosmic proportions because consciousness as an emergent property of the universe would cease to exist and nothing would have any meaning beyond the brute fact of existence.

When poet Richard Wilbur wrote "Advice to the Prophet" at the height of the Cold War, there was a very real possibility that this tragedy would occur after the United States and the Soviet Union unleashed the destructive energy in their nuclear arsenals. According to Wilbur, the prophet who could prevent this horror would arrive on the "streets of our cities / mad-eyed from stating" the "obvious" scientific truths about nuclear war. One piece of advice Wilbur gives his prophet is to "spare us all word of the weapons, their force and range" because we are "unable to fear what is too strange." His rationale for giving this advice is that we cannot conceive of a place where nothing would be aware of its own awareness and the beauty and wonder of the natural world would cease to exist: "How should we dream of this place without us?— / the sun mere fire, the leaves untroubled about us, / a stone look on the stone's face?" Wilbur then tells the prophet that if he can entice people on the streets of our cities to renew their sense of the sacredness, sanctity, and beauty of life, the dream of the earth with us can be realized.

Environmental scientists are now mad-eyed from stating obvious scientific truths about the threats associated with climate change. And many of them are convinced that our species could be on the path toward extinction if global warming triggers irreversible large-scale changes in the climate system. These modern-day Cassandras are also painfully aware that their scientific truths have not been heard by the people on the streets of our cities. And one of primary reasons why this is the case is that the people on these streets, like those in Wilbur's poem, are "unable to fear what is too strange" and cannot "dream of this place without us."

Not surprisingly, these scientists are no longer reluctant to admit that the fate of the earth is sealed by the ignorance, lack of compassion, and inexhaustible greed of its human inhabitants. And since they are reminded on a daily basis that human life could soon become little more than a brutal struggle for survival, all of them are extremely anxious, and most seem to have one or more symptoms of clinical depression. But what my scientific colleagues have apparently failed to realize is that it is possible to dream of this place with us.

If we manage to realize this dream, we could soon find ourselves living in a much more humane and equitable world. In this world, human populations would cease over time to suffer from the ravages of hunger, starvation, and disease; extreme and hopeless poverty would no longer be tolerated or ignored; the rights and freedoms now enjoyed by the citizens of fully functional democracies would be extended to the rest of humanity; and cross-border conflicts and wars of aggression would be viewed as remnants of a benighted human past that have no place or function in the vastly more enlightened human future. The reason why the realization of this dream would result in this more humane and equitable world is obvious in spite of the fact that it is rarely mentioned in the public debate about the environmental crisis: if there is not a very active commitment to creating this world, it will not be possible to achieve the totally unprecedented level of goodwill and cooperation between all peoples and governments required to resolve this crisis.

Realists and pragmatists have routinely dismissed dreams of a better world as the products of the overheated imaginations of muddle-headed idealists who fail to recognize or properly understand what the critics regard as self-evident truths—the more invidious aspects of human nature, the harsh realities of geopolitical politics, the intimate connection between economic prosperity and competitive advantage, the vital role and function of a strong military in protecting and enhancing the interests of sovereign nation-states, and so on. The problem with this conventional wisdom is that the usual distinctions between idealism and realism, moral actions and pragmatic solutions, and idealistic conceptions of the better world that could be versus rational assessments of the world as it is are no longer commensurate with the terms of human survival. What these terms dictate is that idealism is the new realism, moral actions are the only means of achieving pragmatic solutions, and idealistic conceptions of the better world are the only rational basis for preserving and protecting the natural resources that sustain human life.

But for now obvious reasons, realizing the dream of this place will require massive changes in our political and economic institutions and new standards for moral and ethical behavior. And there is a large and growing consensus in the environmental community this can be accomplished only by a well-organized and highly effective worldwide movement in religious environmentalism. The existing worldwide movement has been very effective in dealing with environmental problems on the local or regional level. But it has not made any substantive contributions to the resolution of environmental problems on the global level because it lacks the organization, infrastructure, and resources required to make those contributions. Another, related problem is that the views of nature and the understanding of how environmental problems can or should be resolved are not the same in the great religious traditions of world.

One aim of this chapter is to demonstrate that the new dialogue between the truths of science and religion provides a compelling and coherent basis for resolving these problems. The other is to make the case that if a sufficient number of people in the great religious traditions of the world enter into this dialogue, this could result in the fairly rapid emergence of a worldwide movement in religious environmentalism that, in my view, will be critically important in the effort to prevent the most disastrous consequences of global warming and to resolve other menacing environmental problems. But the price that must be paid to enter this dialogue is a willingness to recognize, as Thomas Berry put it, that "existing religious traditions are too distant from our new sense of the universe to be adequate to the task before us. We cannot do without the traditional religions, but they cannot do what needs to be done." If we are to do what needs to be done, this will require, said Berry, a "new type of religious orientation" that is commensurate with and emerges from "our new story of the universe."[2]

THE NEW STORY OF THE UNIVERSE

In this new story, the cosmos is a seamlessly integrated whole on the quantum mechanical level that evolved toward higher levels of complexity through the process of emergence. As the results of the experiments testing Bell's theorem attest, we live in a nonlocal universe in which quanta can interact with each other over any instance in no time, and any point in space-time is in some sense all points in space-time. In this universe, everything in physical reality is quite literally connected with everything else, and any sense we may have that the collection of particles we call self is isolated and alone is an illusion fostered by a lack of understanding of the actual character of this reality. The results of the experiments testing Bell's theorem also revealed that wave–particle dualism and quantum indeterminacy are indelible features of the life of nature and that there can be no one-to-one correspondence between every element of a physical theory and physical reality. And this means that the language of mathematical physics will never be able to fully disclose and describe this reality.

In the new story about the universe in the biological sciences, life on the blue planet is also a seamlessly interconnected whole that evolved toward higher levels of complexity through the process of emergence. The system of life in this story is self-organizing on the molecular, cellular, and embryological levels, and interactions between organisms regulate the levels of atmospheric gases and sustain conditions that perpetuate the existence of the whole. Feedback loops between proteins, lipids, nucleic acids, cells, tissues, organs, and organisms modify the structures and functions of ecosystems in response to changing conditions in the environment. And this story has also revealed that there is a macro-level indeterminacy in biological reality and that we will never be able to fully disclose or describe this reality in mathematical language or fully understand it in scientific terms.

In the new story about human evolution, all of the seven billion people living on this planet today are the direct descendents of about two thousand individuals in the small tribe of hominids that evolved the capacity to acquire and use fully complex language systems. We now know that the members of the original family of fully modern humans spoke the same language, lived in the same culture, and closely resembled each other in physical terms. We also know that after groups of their descendents migrated out of Africa, language systems and cultures become more diverse, and minor mutations occurred that resulted in variations in skin color, facial features, and body types. In the religious narratives our ancestors created in an effort to coordinate collective activities in the increasingly larger social systems that emerged during and after the agricultural revolution in the Fertile Crescent, the sameness of people in one group was almost invariably defined in terms of perceived differences from people in other groups. But in the new story of science, race and ethnicity have no scientific validity and all members of the extended human family are very similar to one another in genetic, cognitive, and behavioral terms.

The new story about how our ancestors evolved the capacity to acquire and use fully complex language systems has resulted also in a dramatically different understanding of human nature. Research in both the hard and behavioral sciences has revealed that there are evolved and innate neuronal processes in the human brain that result in spontaneous moral behavior on a precognitive level in the absence of feedbacks from higher level cognitive functions associated with the process of making moral decisions. And this research has also revealed that there are evolved and innate capacities in the human brain that make it possible to experience the other as oneself on the precognitive level in the absence of feedbacks from neuronal systems and processes associated with deliberate conscious behavior. In this story, the capacity to engage in spontaneous moral behavior and to experience the other as oneself are aspects of human nature that are from nature and not a product of learning moral norms and values in diverse cultural contexts.

THE TRUTHS OF SCIENCE AND RELIGION

The new story of science has also provided a basis for answering a question that has puzzled students of comparative religion for a long time: why are descriptions of spiritual reality in the sacred texts of the great religious traditions remarkably similar in spite of differences in ontology, or in conceptions of the nature of God or the Ground of Being? In one of the sacred texts of Hinduism, the *Bhagavad Gita*, Lord Krisna says to Arjuna, "Of all that is material and all that is spiritual in this world, know for certain I am both its origins and its dissolution" (Gita 10.8). In the principal sacred text of Taoism, the *Tao te Ching*, the Tao is described as the "Oneness" that cannot be named: "The Above of this Oneness is not bright, and the Below of this Oneness is not dark. As a continuum it cannot be named, and it again vanishes into no-thing-ness. It is therefore the form of no form, the image of no object" (*Tao Te Ching*, 25). In Buddhism, as Ch'an Master Huang Po puts it, "All the Buddhas and all sentient beings are nothing but the One Mind, besides which nothing exists. This Mind, which is without beginning, is unborn and indestructible.... It does not belong to the categories of things which exist and do not exist, nor can it be thought of in terms of new or old."[3]

In the Book of Revelation, God says, "I am the Alpha and Omega, the first and the last" (Rev. 1:8). In the Psalms, "the heavens tell of the glory of God and the firmament bespeak his handwork" (Ps.

19:1), and the "sea and everything in it" sings praises to God (Ps. 96:11). In the New Testament, Christ says, "That they may be one; as thou Father are in me, and I in thee, that they may also be one in us: that the world may believe that thou has sent me. And the Glory which thou has given me I have given to them; that they may be one, even as we are one: I in them and thou in me, that they may be made perfect in one" (John 17:21–23). And it is written in the sacred text of Islam, the *Qur'an*, that, "Whithersoever you turn, there is the face of God" (2:115).

Students of comparative religion have been puzzled also by the fact that descriptions of the most profound religious experience are remarkably similar regardless of the time, place, or cultural context in which they occur. This experience is described as a wordless sense of communion with a single significant whole in which the sense of self as a separate and discrete entity ceases to exist and there is no passage of time or extension in space. According to those who have had this experience, this single significant whole is both infinite and undifferentiated and utterly real in spite of the fact that it cannot be described in cognitive terms.

What the new story of science has to say about this state of profound spiritual awareness involves the results of recent experiments in neuroscience using computer-based brain-imaging systems. In one of these experiments, scientists at the University of Pennsylvania asked Buddhist monks and Catholic nuns to mediate or pray inside a single photon emission tomography (SPEC) brain-imaging system and to signal when they were in a state of profound spiritual awareness by tugging on string. When the monks and nuns indicated that they were in this state, the brain scans showed a dramatic decrease in neuronal activity in the left or dominant hemisphere that contains the major language centers and generates linguistically based narratives about a self separate from world. The scans also showed a large decrease in neuronal activity in the posterior parietal lobes associated with the differentiation and orientation of the body in three-dimensional space and the sense that events are progressing in time.[4]

Since the brain scans of the Buddhist monks and Catholic nuns were virtually identical, this strongly suggests that they were in very similar states of profound spiritual awareness. But when the subjects in these experiments were asked to describe this experience, the monks said they were one with the seamlessly interconnected Ground of Being, and the nuns said they were in the presence of a transcendent and immutable God. The most reasonable explanation for the seeming disparity between the experience of profound spiritual awareness and the descriptions of this experience is that the neurological processes involved are quite different. When the monks and nuns sensed they were one with a single significant whole in spiritual reality, the cognitive processes associated with linguistically based constructions of reality and the experience of self as separate from the world were deactivated. But when they were asked to describe this experience, these processes were reactivated, and the descriptions were based on narratives from their respective religious traditions about the character of spiritual reality and the relationship of self to this reality.

It seems reasonable to conclude that the results of these experiments have revealed that the most profound states of spiritual awareness are now and always have been very similar. If this is the case, it also seems reasonable to conclude that this experience, as nineteenth century psychologist and philosopher William James put it, is the "mother sea and fountain head" of all of the great religious traditions of the world.[5] This would explain not only why descriptions of the most profound state of spiritual awareness are remarkably similar among the great world religions but also why the most profound moral truth in these traditions is that the other is oneself and must be treated as one would wish to be treated.

Hinduism: "One should never do that to another which one regards as injurious to one's own self. This, in brief, is the rule of dharma." Taoism: "Regard your neighbor's gain as your own gain and your neighbor's loss as your own loss." Buddhism: "Hurt not others in ways that you yourself would find hurtful." Judaism: "The stranger who resides with you shall be to you as one of your citizens; you shall love him as yourself." Christianity: "All things whatsoever ye would that men should do to you, do ye even so to them." Islam: "None of you believes until he wishes for his brother what he wishes for himself."[6]

As philosopher Roger Gottlieb points out, religions "are rich, complicated, and multifaceted cultural creations, only parts of which, and not necessarily the most practically important parts, have to do with God."[7] But in spite of the differences in the narratives of the major religious traditions of the world, the most profound religious and moral truths are virtually identical. And as many scholars in comparative religion have demonstrated, the differences can be understood in terms of the diverse environmental conditions and cultural and historical contexts in which these narratives emerged.

This does not mean that we should cease to recognize and celebrate the rich diversity, perennial wisdom, and profound moral insights of the great religious traditions of the world. It does mean, though, that those who are unwilling or unable to accept the fact that the most profound truths in all of these traditions are remarkably similar cannot enter the new dialogue between the truths of science and religion. It also means that a well-organized and highly effective worldwide movement in religious environmentalism will not emerge if spiritually aware and environmentally concerned people are so preoccupied with the differences that they are unable to recognize the sameness.

THE NEW ENVIRONMENTAL ETHOS

The reason why the new dialogue between the truths of science and religion will in my view be critically important in the effort to resolve the environmental crisis is that it provides a basis for articulating and disseminating an environmental ethos with a profound spiritual dimension. In this ethos, all aspects of physical and spiritual reality, including human life and consciousness, are emergent from and embedded in a single significant whole. Those who embrace this ethos will discover that it is possible, as biologist E. O. Wilson put it, "to love life enough to save it."[8] And they will view love, in the words of social psychologist Erich Fromm, as an art that requires "discipline, concentration and patience."[9]

The practitioners of this art will also know that it is necessary, as ecologist Wendell Berry said, to "daily shed the blood and break the body of creation" to sustain human life. But they will do so "knowingly, lovingly, skillfully, reverently" and not "ignorantly, greedily, clumsily, and destructively."[10] In their experience of the natural world, those who embrace this ethos will also realize, as naturalist and essayist John Burroughs puts it, that they are in contact with "primal sanities, primal honesties, primal attractions" and "touching at least the hem of the garment with which the infinite is clothed."[11] And, they will view toxic waste dumps, polluted rivers and streams, dead zones in oceans and estuaries, disseminated rain forests, and all the other wounds that humankind has inflicted on this world as utterly immoral and quite profane.

The spiritually aware and environmentally concerned people who embrace this ethos will also be familiar with what research in neuroscience has revealed about the role of emotions and preconscious and unconscious processes in human thought and behavior. They will know that the cognitive processes associated with empathy, sympathy, and compassion operate largely outside of those associated with linguistically based narratives about self and world. And they will be aware that the primary determinant of our willingness to engage in moral behavior is these feelings and not the moral codes embedded in these narratives.

As neuroscientist Michael Gazzaniga points out, this explains why research in the social sciences has shown that there is little or no "correlation between moral reasoning and proactive moral behavior" and why most studies have not found any correlations.[12] And it also explains, as Rachel Carson, founder of modern environmentalism, put it, why "it is not half so important to know as to feel" and why a sense of wonder in our relationship to the natural world must be experienced and cannot be taught.[13] It is this experience, as Aldo Leopold, one of the fathers of wildlife ecology, famously said, that makes us realize, "A thing is right when it tends to preserve the integrity, stability, and beauty of the biotic community. It is wrong when it tends otherwise."[14]

If we are to realize the dream of this world with us, those who enter the new dialogue between the truths of science and religion must become the prophets who will soon come to the streets of our cities prepared and willing to wage a war aptly described by William James: "What we need to discover in the social realm is the moral equivalent of war: something heroic that will speak to men as universally as war does, and yet will be compatible with their spiritual selves as war proved itself to be incompatible."[15] The prophets in the volunteer army that wages the moral equivalent of war will be citizens of the world who refuse to define themselves in terms of cultural narratives about nationalism and national identity. And the "something heroic" that will be "compatible with their spiritual selves" will be to extend the circle of human compassion to all members of the extended human family and to preserve and protect the environmental resources that sustain human life.

The strategies used in fighting the battles in this war will be based on the philosophy of nonviolence as it was articulated and applied by Mahatma Gandhi and Martin Luther King Jr. The weapons used will be protests, rallies, town meetings, boycotts, and political campaigns promoted and organized with videos, documentaries, films, and web-based communication networks. And the prophets will use these weapons to do battle with the climate change contrarians and the think tanks, special interest groups, corporations, and lobbying firms that fund their highly successful campaign against the truths of science.

Some have argued that the scientific prophets should challenge the big Orwellian lies told by the contrarians in print and electronic media and in well-publicized public debates. But what those who make these arguments have apparently failed to realize is that this would represent a violation of the rules and procedures of doing science. And they also seem unaware of the fact that scientists who violate these rules and procedures would no longer be welcomed at scientific meetings and conferences, their reputations would be badly damaged, their job security would be threatened, and their careers could be effectively over.

The scientific prophets are like messengers who stand at the gate of a city warning its inhabitants that a terrible natural disaster of their own making is looming on the horizon. And as these scientists know very well, they speak scientific truths that the inhabitants of the city often fail to comprehend or choose to ignore. But the scientific truths would no longer be regarded as truth if the messengers entered the city, became members of the polis, and engaged the nonbelievers in well-publicized public debates. One problem here is that the nonbelievers would assume that the truths spoken by the messengers are motivated by political and ideological agendas. Another is that scientists do not have the background or training required to win debates with those who assume that beliefs and opinions have more currency than scientific knowledge and empirically valid facts.

This means that the prophets who must soon come to the streets of our cities must be sufficiently familiar with the truths spoken by the scientific prophets to translate them into language that people on these streets can readily understand. But this language, in contrast with that used by the scientific prophets, will be infused with profound spiritual and moral truths that are recognized as such in all of the great religious traditions of the world. And the prophets will speak these truths in ways that motivate people on the streets of our cities to commit themselves to realizing the dream of this world with us by becoming combatants in a war that is compatible with their spiritual selves.

The prophets who wage these battles will realize, as environmentalist David Orr puts it, that there "are legitimate grounds for hope in hard times, but not one speck of ground for wishful thinking. We won't be rescued by more research, hyper-technology, or some deus ex machina. There is no anonymous 'they' who will figure things out."[16] They will also understand what the late Czech playwright and president Václav Havel meant when he said, "Hope is definitely not the same as optimism. It is not the conviction that things will turn out well, but the certainty that something makes sense, regardless of how it turns out."[17] And the prophets will keep this hope alive by doing good work, rising above the lesser self, and doing the right thing in a spirit of gratitude for and celebration of the gift of life.

The greatest enemy that the prophets will confront in fighting the battles in the war that is compatible with their spiritual selves is not greed and selfishness or any other morally invidious aspects of human behavior. It is the widespread acceptance of assumptions about the sources of human identity and the relationship between self and world in the old stories about political and economic reality. But the prophets will be armed with the knowledge that the new story of science has revealed that narratives about nationalism and national identity are predicated on unscientific assumptions about the sources of human identity. And they will also know that these narratives serve ideological agendas, abuse the truths of religion, perpetuate belief in the existence of false gods, and reinforce this belief in religions of state dedicated to the worship of these gods.

FALSE GODS AND THE POLITICS OF GLOBAL WARMING

In order to better appreciate how menacing this problem is, let us conduct a brief thought experiment. Imagine that scientists as NASA published a report revealing that observational data from satellites in deep space clearly indicates that an asteroid similar in size to that which resulted in a mass extinction 65.5 million years ago will impact the earth in five years. Now suppose that a group of international scientists and engineers devise a $10 trillion plan to prevent this catastrophe that involves the rapid construction and timely launch of precision-built rockets with conventional warheads that would explode in the vicinity of the asteroid and alter its trajectory. Also assume that the process of implementing this plan during the time frame in which it will be possible to divert the path of the asteroid will require a totally unprecedented level of goodwill and cooperation between all peoples and governments.

In this situation, the NASA report would be front page news, and its contents would be endlessly described and discussed in both print and electronic media. Political leaders and economic planners would cease to do business as usual and get down to the business of preventing an ecological disaster that would imperil the existence of humanity. And it is also seems reasonable to assume that the members of the international community would overcome their usual differences and achieve the unprecedented level of goodwill and cooperation required to fund and implement the $10 trillion plan.

One question that this thought experiment is intended to raise is why this has not been the response to predictions about the environmental impacts of global warming made by scientists at NASA and every other major research facility on climate change. There is now an abundance of scientific research which clearly demonstrates that if concentrations of carbon dioxide in the atmosphere continue to increase, there is a high probability that global warming will trigger irreversible large-scale changes in the climate system. And if these changes occur, the environmental impacts would eventually be comparable to those caused by the six-mile-wide asteroid that collided with earth 65.5 million years ago.

Another, related question that the thought experiment is intended to raise is why virtually nothing was said in the mainstream news media about a recent report by an organization that provides decision makers in the fossil fuel business with the data required to manage the growth and expansion of this multi-billion-dollar industry. The organization is the International Energy Agency (IEA), the data in the reports of the IEA are viewed as the gold standard for making decisions about the energy future, and the predictions made in these reports about the environmental impacts of global warming tend to be quite conservative. But in the most recent IEA report, *World Energy Outlook 2011*, the ominous and apparently unnewsworthy prediction was that the international community has only five years to reduce worldwide emissions of greenhouse gases to levels where global warming will not trigger irreversible large-scale changes in the climate system.

Based on a careful analysis of worldwide emissions of greenhouse gases from the existing energy infrastructure, the report concludes that these emissions are already at eighty percent of the level where disastrous changes are projected to occur. The report also makes a convincing case that increases in greenhouse gas emissions from "infrastructure lock-in," or from coal-burning plants, buildings, cars, and trucks that are now being built or will soon be created, will reach one hundred percent of this level by 2017.[18] As Fatih Birol, the chief economist of the IEA, put it, "If we don't change directions now on how we use energy, we will end up beyond what scientists tell us is the limit. The door will be closed forever."[19] The report also estimates that the cost of implementing a plan that could prevent this ecological disaster, $10 trillion, is the same as that of implementing the plan to divert the path of the asteroid in our fictional scenario.

During the press conference that followed the release of the *World Energy Outlook 2011* report in November 2011, the directors of the IEA made it quite clear that were issuing a wake-up call to humanity.[20] They also made it clear that this call was being directed at the representatives of 194 sovereign nation-states who in Durban, South Africa, in December 2011, were charged with the task of implementing a post-Kyoto agreement on global warming. But what the directors of the IEA apparently failed to realize is that the goal of these representatives is not to serve humanity. It is to protect and enhance the vested economic interests of particular nation-states and to promote the growth and expansion of the global market system in ways that serve those interests.

The existing agreement on global warming was adopted in Kyoto, Japan, in 1997, and its stated objective was to stabilize concentrations of greenhouse gases in the atmosphere "at a level that would prevent dangerous anthropogenic interference with the climate system."[21] Under the terms of this agreement, thirty-seven developed countries with highly industrialized economies pledged to reduce their emissions of greenhouse gases by 5.2 percent below 1990 levels during the period 2008–2012. But the developed countries did not keep this pledge, and China and India, major developing countries, were not parties to the agreement.

There was some coverage in the mainstream news media of the failed effort by the international community during the summit in Durban to implement a post-Kyoto agreement. And a few passing mentions were made of the fact that the agreement that emerged after two weeks of exhaustive and ill-tempered negations will not stabilize concentrations of greenhouse gases in the atmosphere "at a level that would prevent dangerous anthropocentric interference with the climate system." But to my knowledge nothing was said in the mainstream news media about the fact that the so-called Durban Platform for Enhanced Action agreement will allow these emissions to increase to levels where a great deal of scientific research predicts that dangerous interference with the climate system will occur.

This agreement states that there will be a second round of negotiations after the provisions of the existing agreement on global warming expire in 2012. It also stipulates that the mandatory provisions in the post-Kyoto agreement that will allegedly result from these negotiations will apply to developing countries in spite of the fact that the governments of these countries are adamantly opposed to this idea. The prospects that this agreement will be implemented during a period in which most countries are experiencing economic malaise and political uncertainty are probably slim to none. And even in the unlikely event that this agreement is implemented, the mandatory reductions would not begin until 2020.

The large and now familiar problem here is that this approach to resolving the environmental crisis is a program for ecological disaster predicated on the following list of assumptions which testify to a belief in the existence of false gods: (1) the lawful or law-like dynamics of free market systems can resolve virtually all environmental problems; (2) these dynamics will necessarily result in technological solutions; (3) governments should deal only with those environmental problems that cannot be resolved in these terms; (4) any actions taken by government must be commensurate with the understanding of the dynamics of free market systems in the neoclassical economic paradigm; (5) the sole source of political power in dealing with environmental problems is the sovereign nation-state; (6) the sovereign nation-state exists, like all forms of government, in a domain of reality separate and distinct from the domain of economic reality; (7) the international system of government does not in itself have any political power; and (8) any attempts by this government to resolve environmental problems which might interfere with the growth and expansion of national economies and the global market system must be resisted by the governments of sovereign nation-states.

If the natural laws of economics and the sovereign nation-state were sacredly ordained aspects of a cosmic scheme or plan, business as usual could be very good business indeed. But since these constructs exist only in the minds of those who believe in their existence, and since belief in their existence is effectively undermining efforts to resolve the environmental crisis, clearly the time has come to recognize that living in the service of false gods is not in the interests of human survival. If we are to realize the dream of this world with us, the prophets who must soon enter the streets of our cities must not only challenge belief in the existence of these false gods.

They must also begin to create a well-organized and highly effective worldwide movement in religious environmentalism that is committed to accomplishing two now-familiar objectives. The first is to create a supranational system of federal government that is capable of implementing the scientifically viable public policies and economic programs required to prevent the most disastrous consequences of global warming in a binding international agreement. The second is to replace neoclassical economic theory with an environmentally responsible theory that can serve as the basis for implementing scientifically viable economic solutions for environmental problems.

Fortunately, a well-organized and highly effective worldwide movement in religious environmentalism could be created rather quickly by linking and coordinating activities and initiatives in the existing worldwide movement. The National Religious Partnership for the Environment and the Earth Charter Commission are attempting to wed ecological values to fundamental religious truths in all of the great religions of the world.[22] The Boston-based Tellus Institute and the Forum on Religion and Ecology at Yale also are actively engaged in this effort. And the United Church of Christ has played a decisive role in making environmental justice a critically important issue in the existing worldwide movement of religious environmentalism.

The World Council of Churches has sponsored international conferences on how to achieve socioeconomic justice and ecological sustainability, and the Interfaith Climate Change Network has organized eighteen statewide interfaith climate change campaigns. Groups of Lutherans, Shintoists, and Maronite Christians have been successful in protecting forests in Sweden, Japan, and Lebanon. And religious leaders at Jewish synagogues in England and Buddhists pagodas in Cambodia are preaching and practicing a new green gospel. Equally impressive, Sikhs in India, through their network of 28,000 temples, have provided solar power and fuel-efficient technologies to tens of thousands of impoverished people and initiated a series of projects to raise ecological awareness, reduce pollution, and improve damaged ecosystems.

As Gottlieb points out, "Activist religious environmentalism goes beyond theology and public declarations. It is directly aimed at changing the world by making new laws, stopping harmful practices, creating better ways to produce and consume, healing the earth, and nurturing human beings in their relations with the rest of life."[23] But as Gottlieb knows only too well, most people in the religious environmentalism movement have not gone beyond theology and public declarations. What these spiritually aware and environmentally concerned people will hopefully realize is that making the changes required to resolve the environmental crisis is not merely a moral imperative. It is an existential imperative.

People who worship at the altar of false gods will not appreciate being told that these gods do not exist and that belief in their existence will soon result in a human tragedy of staggering dimensions. But if the prophets extend the circle of their compassion to these true believers and enlarge the bases for mutual recognition and understanding, most should realize that the death of false gods is a small price to pay for a once-in-all-human-lifetimes opportunity. The opportunity is to protect the lives of existing members of the extended human family and the future existence of subsequent generations of this family by resolving the crisis in the global environment.

It is, however, reasonable to assume that many true believers who worship at the altar of false gods in the religion of The Market and in churches of state will be unwilling or unable to embrace this opportunity or even to recognize that it exists. It is also reasonable to assume that they will defend their belief in the existence of these gods on conservative talk shows and news programs, in articles published in magazines and newspapers, in infomercials on radio and television, and in the well-financed campaign of the climate change contrarians. But unlike the scientific prophets, the prophets who wage the war that is compatible with their spiritual selves can enter the gates of our cities, engage the political process, and do battle with the true believers who tell the big Orwellian lies. And the prophets who fight these battles will be armed with the truths that can expose these lies and be prepared and willing to use all nonviolent means available to realize the dream of this place with us by winning this war.

My last two comments are directed to readers who, for professional or personal reasons, may be reluctant to enlist in the volunteer army of environmental activists that will wage this war. The first is that if the dream of this place with us is realized, we will soon find ourselves living a more just and peaceful world in which extreme poverty does not exist and universal rights and freedoms are extended to all of humanity. And the second is that this is not merely the work of an age, but a work that can preserve the memory of all ages, and it is not possible to serve a greater good or answer to a higher calling.

NOTES

INTRODUCTION
1. Nathanial Gronewold, "Pakistan—A Sad New Benchmark in Climate-Related Disasters," *New York Times*, Aug. 18, 2010.
2. Camille Parmesan, "Ecological and Evolutionary Responses to Recent Climate Change," *Annual Review of Ecology, Evolution, and Systematics* 37 (Dec. 2006): 637–639; James E. Hansen et al., "Dangerous Human-Made Interference with Climate: A GISS ModelE Study, *Atmospheric Chemistry and Physics* 7 (May 2007): 2287–2312; http://www.atmos-chem-phys.net/7/2287/2007/acp-7-2287-2007.pdf; Intergovernmental Panel on Climate Change (IPCC), *Climate Change 2007*, ed. Susan Soloman (New York: Cambridge University Press, 2007); U.S. Climate Change Science Program (CCSP), *Final Report of Synthesis and Assessment Product* 4.1, 4.2, 2.3, 1.2 (all posted Jan. 16, 2009), www.usgcrp.gov/usgcrp/default.php.
3. Quoted in Al Gore, "Climate of Denial: Can Science and Truth Withstand the Merchants of Poison?" *Rolling Stone*, June 22, 2011, www.rollingstone.com/politics/news/climate-of-denial-20110622.
4. James E. Hansen et al., "Target Atmospheric CO_2: Where Should Humanity Aim?" *Open Atmospheric Science Journal* 2 (2008): 217–231, available at http://earth.geology.yale.edu/~mp364/data/2008%20Hansen.pdf.
5. See pp. 4–5 of Peter Schwartz and Doug Randall, "An Abrupt Climate Change Scenario and Its Implications for United States National Security," report prepared by the Global Business Network for the U.S. Department of Defense (Oct. 2003), www.s-e-i.org/pentagon_climate_change.pdf.
6. Schwartz and Randall, "An Abrupt Climate Change Scenario and Its Implications for United States National Security," p. 7.
7. Schwartz and Randall, "An Abrupt Climate Change Scenario and Its Implications for United States National Security," p. 2.
8. Schwartz and Randall, "An Abrupt Climate Change Scenario and Its Implications for United States National Security," p. 18.

CHAPTER 1

1. Thomas Berry, *The Dream of the Earth* (San Francisco: Sierra Club Books, 1988), p. 123.
2. "An Open Letter to the Religious Community," in Mary Evelyn Tucker, *Worldly Wonder: Religions Enter Their Ecological Phase* (Chicago: Open Court, 2003), pp. 116–123.
3. Bill McKibben, "Where Do We Go from Here?" *Daedalus* 130 (Fall 2001), www.amacad.org/publications/fall2001/mckibben.aspx.
4. Roger S. Gottlieb, *A Greener Faith: Religious Environmentalism and Our Planet's Future* (New York: Oxford University Press, 2006), p. 151.
5. Gottlieb, *Greener Faith*.
6. Fred Pearce, "Arctic Meltdown Is a Threat to Humanity," *New Scientist*, March 25, 2009.
7. Rockström et al., "A Safe Operating Space for Humanity," *Nature* 461 (Sept. 24, 2009): 472–475.
8. Mason Inman, "Arctic Ice in 'Death Spiral,'" *National Geographic News*, Sept. 7, 2007.
9. Steve Connor, "Expanding Tropics a Threat to Millions," *Independent*, Dec. 3, 2007.
10. Frances C. Moore, "Climate Change and Air Pollution: Exploring the Synergies and Potential for Mitigation in Industrializing Countries," *Sustainability* 1, no.1 (2009): 43–54.
11. Susan Soloman et al., "Irreversible Climate Change due to Carbon Dioxide Emissions," *Proceedings of the National Academy of Sciences* 106, no. 6 (2009): 1704–1709.
12. Zoological Society of London (ZSL), "Coral Reefs Exposed to Imminent Destruction from Climate Change," *Science Daily*, July 6, 2009, www.sciencedaily.com/releases/2009/07/090706141006.htm.
13. Aradhna K. Tripati, Christopher D. Roberts, and Robert A. Eagle, "Coupling of CO_2 and Ice Sheet Stability over Major Climate Transitions of the Last 20 Million Years," *Science* (online), Oct. 8, 2009, www.sciencemag.org/content/326/5958/1394.abstract.
14. James E. Hansen et al., "Target Atmospheric CO_2: Where Should Humanity Aim?" *Open Atmospheric Science Journal* 2 (2008): 217–231, http://earth.geology.yale.edu/~mp364/data/2008%20Hansen.
15. Rockström et al., "A Safe Operating Space for Humanity." p. 473.
16. Camille Parmesan, "Ecological and Evolutionary Responses to Recent Climate Change," *Annual Review of Ecology, Evolution, and Systematics* 37 (Dec. 2006): 637–639; James E. Hansen et al., "Dangerous Human-Made Interference with Climate: A GISS ModelE Study, *Atmospheric Chemistry and Physics* 7 (May 2007): 2287–2312; Intergovernmental Panel on Climate Change (IPCC), *Climate Change 2007*, ed. Susan Soloman (New York: Cambridge University Press, 2007); U.S. Climate Change Science Program (CCSP), *Final Report of Synthesis and Assessment Product 4.1, 4.2, 2.3, 1.2* (all posted Jan. 16, 2009), www.usgcrp.gov/usgcrp/default.php.

17. Frank Lutz quoted in Mitchell Anderson, "Trust Us, We're the Media," straight.com (Jan. 25, 2007), www.straight.com/article-67107/trust-us-were-the-media.
18. Pew Research Center for the People & the Press, "Fewer Americans See Solid Evidence of Global Warming" (Oct. 22, 2009), www.people-press.org/report/556/global-warming.
19. Elisabeth Rosenthal, "Climate Fears Turn to Doubts among Britons," *New York Times*, May 24, 2010, www.nytimes.com/2010/05/25/science/earth/25climate.html.

CHAPTER 2
1. Lynn Jorde, Michael Bamshad, and Alan Rogers, "Using Mitochondrial Nuclear DNA to Reconstruct Human Evolution," *BioEssays* 20 (1998): 126–136; A. Gibbons, "The Mystery of Humanity's Missing Mutations," *Science* 267 (Jan. 6, 1995): 35–36; L. L Cavalli-Sforza, P. Menozzi, and A. Piazza, "Demic Expansions and Human Evolution," *Science* 259 (Jan. 29, 1993): 639–646.
2. Richard G. Klein, *The Human Career: Human Biology and Cultural Origins*, 2nd ed. (Chicago: University of Chicago Press, 1999); Klein, "Archaeology and the Evolution of Human Behavior," *Evolutionary Anthropology* 9, no. 1 (2000): 17–36.
3. John D. Bengtson and Merritt Ruhlen, "Global Etymologies," in *On the Origins of Languages: Studies in Linguistic Taxamony*, ed. Merritt Ruhlen (Stanford, Calif.: Stanford University Press, 1994), pp. 278–336.
4. J. Mehler et al., "A Precursor of Language Acquisition in Young Infants," *Cognition* 29 (July 1988): 143–178.
5. Steven Pinker, *Language Learnability and Language Development* (Cambridge, Mass: Harvard University Press, 1984); Brian MacWhinney, ed., *Mechanisms of Language Acquisition* (Hillsdale, N.J.: Erlbaum, 1987).
6. Bill Bryson, *The Mother Tongue: English and How It Got That Way* (New York: Morrow, 1990).
7. G. A. Miller, *The Science of Words* (New York: Freedman, 1991).
8. Giacomo Rizzolatti quoted in Sandra Blakeslee, "Cells That Read Minds," *New York Times*, Jan. 10, 2006.
9. Seymour Epstein, "Integration of the Cognitive and Psychodynamic Unconscious," *American Psychologist* 49 (Aug. 1994): 710.
10. Paul Slovak, "'If I look at the mass I will never act': Psychic Numbing and Genocide," *Judgment and Decision Making* 2 (April 2007): 79–95.
11. K. E. Jenni and G. Loewenstein, "Explaining the Identifiable Victim Effect," *Journal of Risk and Uncertainty* 14 (1997): 235–257; T. Kohut and I. Ritov, "The Singularity of Identified Victims in Separate and Joint Evaluations," *Journal of Behavioral Decision Making* 18 (2005): 106–116.
12. D. A. Small, G. Loewenstein, and P. Slovak. "Sympathy and Callousness: The Impact of Deliberate Thought on Donations to Identifiable and Statistical Victims," *Organizational Behavior and Human Decision Processes* 102 (2006): 143–153.
13. Annie Dillard, *For the Time Being* (New York: Knopf, 1999), p. 47.

14. Donald Brown, *Human Universals* (New York: McGraw Hill, 1991); Donald Brown, "Human Universals: Human Nature and Human Culture," *Daedalus* 133 (2004): 4–47.
15. Natalie Angier, "Thirst for Fairness May Have Helped Us Survive," *New York Times*, July 4, 2011.
16. Gary Stix, "Traces of a Distant Past," *Scientific American*, July 2008.
17. Steven Molnar, *Human Races, Types, and Ethnic Groups*, 4th ed. (Upper Saddle River, N.J.: Prentice Hall, 1988); Nina Jablonski and George Chaplin, "The Evolution of Human Skin Coloration," *Journal of Human Evolution* 39 (2000): 57–106.
18. Lyle Campbell, *Historical Linguistics* (Cambridge, Mass: MIT Press, 1999).
19. Jared Diamond, *Guns, Germs, and Steel: The Fates of Human Societies* (New York: Norton, 1999), pp. 132–133.
20. E. O. Wilson, *The Future of Life* (New York: Alfred A. Knopf, 2002), pp. 98–99.
21. E. O. Wilson, *Consilience: The Unity of Knowledge* (New York: Knopf, 1998), p. 262.
22. Alfred Crosby, *The Columbian Exchange: Biological and Cultural Consequences of 1492* (Westport, Conn.: Greenwood, 1972).
23. Richard P. Clark, *Global Life Systems* (New York: Rowan & Littlefield, 2000), pp. 165–186.
24. William Cronon, *Changes in the Land: Indians, Colonists, and the Ecology of New England* (New York: Hill & Wang, 1983); Timothy Silver, *A New Face on the Countryside: Indians, Colonists, and Slaves in South Atlantic Forests, 1500–1800* (Cambridge: Cambridge University Press, 1990).
25. Alfred Crosby, *Germs, Seeds and Animals: Studies in Ecological History* (Armonk, N.Y.: Sharpe, 1994), pp. 148–156.
26. Nate Hagens, "Beyond the Barrel," *U.S. News & World Report*, Jan. 7, 2008.
27. Clive Pointing, *A Green History of the World: The Environment and the Collapse of Great Civilizations* (New York: St. Martin's, 1991).
28. William H. McNeill, *Plagues and Peoples* (New York: Doubleday, 1977).
29. Joseph Konvitz, *The Urban Millennium: The City-Building Process from the Early Middle Ages to the Present* (Carbondale: Southern Illinois University, 1985), chap. 4.
30. Herman Prager, "Commentary on U.N. Environment Program," in *Global Marine Environment: Does the Water Planet Have a Future?* (Lanham, Md.: University Press of America, 1993), pp. 61–62.
31. David Orr, *Earth in Mind: On Education, Environment, and the Human Prospect* (Washington, D.C.: Island, 2004), p. 56.
32. Peter Freund and George Martin, *The Ecology of the Automobile* (Montreal: Black Rose Books, 1993).
33. James J. MacKenzie and Michael P. Walsh, *Driving Forces: Motor Vehicle Trends and Their Implications for Global Warming* (Washington, D.C.: World Resources Institute, 1990); "One Billion Cars," *Worldwatch*, Jan.–Feb. 1996.

34. J. R. McNeill, *Something New under the Sun: An Environmental History of the Twentieth Century* (New York: Norton, 2000), p. 311.
35. Robert Ross and Kent Trachte, *Global Capitalism: The New Leviathan* (Albany: State University of New York Press, 1990); William Grieder, *One World, Ready or Not: The Manic Growth of Global Capitalism* (New York: Simon & Schuster, 1997).
36. Brown, *Human Universals*.

CHAPTER 3
1. Henry P. Stapp, "Quantum Theory and the Physicist's Conception of Nature: Philosophical Implications of Bell's Theorem," in *The World View of Contemporary Physics: Does It Need a New Metaphysics?* ed. Richard F. Kitchener (Albany: State University of New York Press, 1988), p. 38.
2. Werner Heisenberg, *Physics and Philosophy* (London: Faber, 1959), p. 96.
3. Max Planck, *Where Is Science Going?* (London: Allen & Unwin, 1933), p. 24.
4. Albert Einstein, "Autobiographical Notes," in *Albert Einstein: Philosopher–Scientist*, ed. P. A. Schlipp (New York: Harper & Row, 1959), p. 3.
5. Albert Einstein quoted in *New York Post*, Nov. 28, 1972.
6. Pierre-Simon de Laplace quoted in Ronald W. Clark, *Einstein: The Life and Times* (New York: World Press, 1971), p. 34.
7. Henry P. Stapp, "Quantum Theory and the Physicist's Conception of Nature: Philosophical Implications of Bell's Theorem," in *The World View of Contemporary Physics: Does It Need a New Metaphysics?* ed. Richard F. Kitchener (Albany: State University of New York Press, 1988), p. 54.
8. Richard Feynmann, *The Character of Physical Law* (Cambridge, Mass: MIT Press, 1967), p. 130.
9. John A. Wheeler, "Beyond the Black Hole," in *Some Strangeness in the Proportion*, ed. Harry Woolf (London: Addison–Wesley, 1980), p. 354.
10. Abner Shimony, "The Reality of the Quantum World," *Scientific American*, Jan. 1988, p. 46.
11. Brian Swimme and Thomas Berry, *The Universe Story* (San Francisco: Harper Collins, 1992), p. 27.
12. Alain Aspect, Jean Dalibard, and Gérard Roger, "Experimental Test of Bell's Inequalities Using Time-Varying Analyzers," *Physical Review Letters* 49 (Dec. 20, 1982): 1804.
13. W. Tittle et al., "Violation of Bell's Inequalities by Photons More Than 10km Apart," *Physical Review Letters* 81 (Oct. 26, 1988): 3563–3566.
14. One of the gross misinterpretations in the popular press of the results of the experiments testing Bell's theorem was that they showed that information traveled between the detectors at speeds greater than light. This was not the case, and relativity theory, along with the rule that the velocity of light is the absolute limit at which signals can travel, was not violated. The proper way to view these correlations is that they occurred instantly or in "no time" in spite of the vast distance between the detectors. The results also indicate that similar correlations would occur even if the distance between the detectors was billions of light years.

A number of articles in the popular press also claimed that the results of the Gisin experiments showed that faster-than-light communication is possible. This misunderstanding resulted from a failure to appreciate the fact there is no way to carry useful information between paired particles in this situation. The effect that is studied in these experiments applies only to events that have a common origin in a unified quantum system, like the annihilation of a positron–electron pair, the return of an electron to its ground state, or the separation of a pair of photons from the singlet state. Since any information that originates from these sources involves quantum indeterminacy, the individual signals are random, and random signals cannot carry coded information or data.

The polarizations, or spins, of each of the photons in the Gisin experiments carry no information, and any observer of the photons transmitted along a particular axis would see only a random pattern. This pattern makes nonrandom sense only if we are able to compare it with the pattern observed in the other paired photon. Any information contained in the paired photons derives from the fact that the properties of the two photons exist in complementary relation, and that information is uncovered only through a comparison of differences between the two random patterns.

15. Bernard d'Espagnat, *In Search of Reality* (New York: Springer–Verlag, 1981), pp. 43–48.
16. N. David Mermin, "Extreme Quantum Entanglement in a Superposition of Macroscopically Distinct States," *Physical Review Letters* 65 (Oct. 8, 1990): 1838–1840.
17. Robert Nadeau and Menas Kafatos, *The Non-Local Universe: The New Physics and Matters of the Mind* (New York: Oxford University Press, 1999).
18. Erwin Schrodinger in *Quantum Questions: Mystical Writings of the World's Greatest Physicists*, ed. Ken Wilbur (Boston: Shambhala, 1984), p. 97.
19. Freeman Dyson quoted in *Nature's Imagination: The Frontiers of Scientific Vision*, ed. John Cornwell (Oxford: Oxford University Press, 1995), p. 8.
20. Rudy Rucker, *Infinity and Mind* (Boston: Birkhauser, 1982), p, 157.
21. Schrodinger in *Quantum Questions*, p. 81.
22. Wolfgang Pauli in *Quantum Questions*, p. 163.

CHAPTER 4
1. Charles Darwin, "The Linnean Society Papers," in *Darwin: A Norton Critical Edition*, ed. Philip Appleman (New York: Norton, 1970), p. 83.
2. Charles Darwin, *The Origin of Species* (New York: Mentor, 1958), p. 75
3. Ernst Mayr, *The Growth of Biological Thought: Diversity, Evolution, and Inheritance* (Cambridge, Mass.: Harvard University Press, 1982), p. 63.
4. P. B. Medawar and J. S. Medawar, *The Life Sciences: Current Ideas in Biology* (New York: Harper & Row, 1977), p. 165.
5. Harold Morowitz, *Beginnings of Cellular Life* (New Haven, Conn.: Yale University Press, 1992); Morowitz, *The Emergence of Everything* (New York: Oxford University Press, 2002).

6. Lynn Margulis and Dorion Sagan, *Microcosmos: Four Billion Years from Our Microbial Ancestors* (New York: Simon & Schuster, 1986), p. 18.
7. Margulis and Sagan, *Microcosmos,* p 18.
8. Margulis and Sagan, *Microcosmos,* p. 19
9. Margulis and Sagan, *Microscomos,* p. 265.
10. Marjori Matzke and Antonius J. M. Matzke, "RNAi Extends Its Reach," *Science* 391 (Aug. 22, 2003): 1060–1061.
11. Darwin, *Origin of Species*, p. 83.
12. Darwin, *Origin of Species*, p. 77.
13. Darwin, *Origin of Species*, p. 75.
14. Darwin, *Origin of Species*, p. 76.
15. Darwin, *Origin of Species*, pp. 78–79.
16. Richard M. Laws, "Experiences in the Study of Large Animals," in *Dynamics of Large Mammal Populations*, ed. Charles W. Fowler and Tim D. Smith (New York: Wiley, 1981), p. 27.
17. Charles Fowler, "Comparative Population Dynamics in Large Animals," in *Dynamics of Large Mammal Populations*, pp. 444–445.
18. Charles Elton, *Animal Ecology* (London: Methuen, 1968), p. 119.
19. James L. Gould, *Ethology: Mechanisms and Evolution of Behavior* (New York: Norton, 1982), p. 467.
20. Paul Colinvaux, *Why Big Fierce Animals Are Rare: An Ecologist's Perspective* (Princeton, N.J.: Princeton University Press, 1978), p. 145.
21. Colinvaux, *Why Big Fierce Animals Are Rare*, p. 146.
22. Peter Farb, *The Forest* (New York: Time Life, 1969), p. 116.
23. Peter H. Klopfer, *Habitats and Territories* (New York: Basic Books, 1969), p. 9.
24. Eugene P. Odum, *Fundamentals of Ecology* (Philadelphia: Saunders, 1971), p. 216.
25. J. Shaxel, *Gruduz der Theorienbuldung in der Biologie* (Jena: Fisher, 1922), p. 308.
26. Natalie Angier, "Constantly in Motion, Like DNA Itself," *New York Times*, March 2, 2004.
27. Errol E. Harris, "Contemporary Physics and Dialectical Holism," in *The World View of Contemporary Physics: Does It Need a New Metaphysics?* ed. Richard F. Kitchener (Albany: State University of New York Press, 1988), p. 159.
28. Werner Heisenberg in *Quantum Questions: Mystical Writings of the World's Greatest Physicists*, ed. Ken Wilbur (Boston: Shambhala, 1984), p. 96.

CHAPTER 5
1. Hagen Schulze, *States, Nations, and Nationalism: From the Middle Ages to the Present,* trans. William E. Yuill (London: Blackwell, 1998), pp. 8–10.
2. George Mosse, *Confronting the Nation: Jewish and Western Nationalism* (Hanover, N.H.: University Press of New England, 1994).
3. Anthony D. Smith, *Myths and Memories of the Nation* (New York: Oxford University Press, 1999), pp. 65–66.

4. Niccolò Machiavelli, *Reflections on the First Ten Books of Titus Livius*, bk. 1, chap. 3, p. 17.
5. Arthur L. Stinchcombe, "Social Structures and Politics," in *Handbook of Political Science*, vol. 3, ed. Fred I. Greenstein and Nelson W. Polsby (Reading, Mass.: Addison–Wesley, 1975), pp. 600–601.
6. Michael Billig and Henry Tajfel, "Social Categorization and Similarity in Intergroup Behavior," *European Journal of Social Psychology* 2, no. 1 (1973): 27–52; Henry Tajfel, "Experiments in Intergroup Discrimination," *Scientific American*, Nov. 1970, pp. 96–102; Henry Tajfel and John C. Turner, "The Social Identity Theory of Intergroup Behavior," in *Psychology of Intergroup Relations*, ed. Stephen Worchel and W. G. Austin (Chicago: Nelson–Hall, 1986), pp. 7–16; Henry Tajfel, *Differentiation between Social Groups* (London: Academic Press, 1972).
7. Sigmund Freud, *Civilization and Its Discontents* (New York: Norton, 1961), p. 61.
8. Pope Leo X quoted in Jackson J. Spielvogel, *Western Civilization* (Stamford, Conn.: Wadsworth, 2002), p. 265.
9. Schulze, *States, Nations, and Nationalism*, p. 164.
10. Heinrich Luden, *Geschichte des Teutschen Volkes*, 12 vols. (Berlin: Gotha, 1825–1829).
11. Immanuel Wallerstein, *The Modern World-System* (New York: Academic Press, 1974), p. 207.
12. Peter Singer, *One World: The Ethics of Globalization* (New Haven, Conn.: Yale University Press, 2002), p. 146.
13. Micheal M'Gonigle and Mark W. Zacher, *Pollution, Politics, and International Law* (Berkeley: University of California Press, 1979).
14. James Gustave Speth, *Red Sky at Morning: America and the Crisis in the Global Environment* (New Haven, Conn.: Yale University Press, 2004), pp. 77–98.
15. The following is an interesting case study on the manner in which scientific evidence was explicitly rejected as the basis for decision making on ocean dumping of radioactive wastes: Judith Spiller and Cynthia Haden, "Radwaste at Sea: A New Era of Polarization or a New Basis for Consensus?" *Ocean Development and International Law* 19, no. 5 (1988): 345–366.
16. Uwe Buse, "Is the IPCC Doing Harm to Science?" *Spiegel Online*, May 3, 2007, www.spiegel.de/international/world/0,1518,480766,00.html.
17. Juliet Eilperin, "U.S. Trying to Weaken G-8 Climate Change Declaration," *Boston Globe*, May 14, 2007.
18. Thomas Fuller and Andrew C. Rifkin, "Climate Plan Looks beyond Bush's Tenure," *New York Times*, Dec. 16, 2007.
19. Joseph Coleman, "Bali Talks Reach Agreement," Associated Press, Dec. 16, 2007.
20. "Disappointment on Climate," *New York Times*, Nov. 17, 2007.
21. Christine Todd Whitman quoted in Erin Pianin, "U.S. Rebuffs Europeans Urging Change of Mind on Kyoto Treaty," *Washington Post*, April 4, 2001.
22. George W. Bush quoted in Edmund L. Andrews, "Bush Angers Europe by Eroding Pact on Warming," *New York Times*, April 1, 2001.

23. Romano Prodi quoted in Andrews, "Bush Angers Europe by Eroding Pact on Warming."
24. D. L. Horowitz, *The Deadly Ethnic Riot* (Berkeley: University of California Press, 2001); John Keegan, *The Face of Battle* (New York: Penguin, 1976).
25. Frank R. Baumgartner and Bryan D. Jones, *Agenda and Instability in American Politics* (Chicago: University of Chicago Press, 1993).

CHAPTER 6

1. Robert Nadeau, *The Wealth of Nature: How Mainstream Economics Has Failed the Environment* (New York: Columbia University Press, 2003), pp. 19–36; Nadeau, *The Environmental Endgame: Mainstream Economics, Ecological Disaster, and Human Survival* (New Brunswick, N.J.: Rutgers University Press, 2006), pp. 102–123.
2. K. J. Arrow and F. H. Hahn, *General Competitive Analysis* (San Francisco: Holden Day, 1971), p 1.
3. Robert Heilbroner, *The Worldly Philosophers; The Lives, Times, and Ideas of the Great Economic Thinkers* (New York: Simon & Schuster, 1992), pp. 42–43.
4. Hielbroner, *Worldly Philosophers*, pp. 33–50.
5. Adam Smith, *An Inquiry into the Nature and Causes of the Wealth of Nations*, ed. R. H. Campbell, A. S. Skinner, and W. B. Todd (Oxford: Oxford University Press, 1976), Astronomy intro. 1.7.II.2.
6. Smith, *Wealth of Nations*, III.2.
7. Smith, *Wealth of Nations*, Astronomy, II.12; III.3.
8. Smith, *Wealth of Nations*, Astronomy, II.12; III.3.
9. Smith, *Wealth of Nations*, IV.ix.51.
10. Smith, *Wealth of Nations*, V.i.f.28.
11. Smith, *Wealth of Nations*, Physics 9.
12. Smith, *Wealth of Nations*, Physics 9.
13. Adam Smith, *The Theory of Moral Sentiments*, ed. D. D. Raphael and A. L. Macfie (Oxford: Oxford University Press, 1976), IV.1.10.
14. Smith, *Theory of Moral Sentiments*, IV.1.10.
15. Smith, *Wealth of Nations*, Astronomy IV.19.
16. Peter Minowitz, *Profits, Priests, and Princes: Adam Smith's Emancipation of Economics from Politics and Religion* (Stanford, Calif.: Stanford University Press, 1993), p. 131.
17. Smith, *Theory of Moral Sentiments*, VII.ii.1.20.
18. Smith, *Theory of Moral Sentiments*, I.i.4.2.
19. Smith, *Wealth of Nations*, VI.i.11–12.
20. Smith, *Wealth of Nations*, VI.i.13.
21. Smith, *Theory of Moral Sentiments*, IV.I.10.
22. Smith, *Wealth of Nations*, VI.I.14.
23. William Godwin quoted in James Bonar, *Malthus and His Work* (New York: Augustus Kelly, 1967) p. 15.
24. Thomas Robert Malthus, *An Essay on the Principle of Population*, ed. Philip Appelman (New York: Norton, 1976), pp. 15ff.
25. Malthus, *Essay on the Principle of Population*.

26. Heilbroner, *Worldly Philosophers*, p. 95.
27. David Ricardo, "On Rent," in Robert Hielbroner, *Teachings from the Worldly Philosopers* (New York: Norton, 1996), p. 116.

CHAPTER 7
1. Philip Mirowski, *Against Mechanism: Protecting Economics from Science* (Lanham, Md.: Rowman & Littlefield, 1988).
2. Morowsky, *Against Mechanism*, pp. 19–20.
3. For readers interested in a detailed discussion of the manner in which the creators of neoclassical economics abused mid–nineteenth century physics, the best available source is Morowski's *Against Mechanism*. The following is a less robust treatment that illustrates how these economists appropriated the mathematics of this physics and redefined energy as the equivalent of utility.

    Assume a mass point is displaced from point A to B in a three-dimensional plane by force vector F and that the force vector is decomposed into its perpendicular components, $F = iF_x + jF_y + kF_z$, where the notation i, j, k represents unit vectors along the three spatial axes. In the same manner, assume that the vector of displacement dq can also be decomposed into its perpendicular components, $dq = id_x + jd_y + kd_z$. Hence the work accomplished, or the product of the force and the infinitesimal displacements, is defined as the integral of the force times the displacement:

    $$T = \int_A^B \left(F_x d_x + F_y d_y + F_z d_z \right) = \frac{1}{2}mv^2 \,|\, B - \frac{1}{2}mv^2 \,|\, A$$

    The mid–nineteenth century physicists redefined the change in mv as the change in the kinetic energy of the particle and represented this as a single-value vector function with T representing the change in kinetic energy. Assume that $(F_x d_x + F_y d_y + F_z d_z)$ is an exact differential and that there exists a uniquely identified scalar function $U(x, y, z)$ such that

    $$F_x = -\partial U/\partial_x;\ F_y = -\partial U/\partial_y;\ F_z = -\partial U/\partial_y.$$

    The scalar function U was viewed as the unobserved potential energy of the particle, and the total energy of the particle, which is presumably conserved through any motion, was represented as $T + U$. William Hamilton had earlier defined the action integral over time of the path of the particle as

    $$\int_{t,A}^{t,B} (T - U)\,dt$$

    The Hamiltonian principle of least action asserts that the actual path of the particle from A to B will be the one that makes the action interval stationary, and this path can be calculated by finding the constrained extrema using either Langrangean constrained maximization/minimization techniques or the calculus of variations. In a conservative system where $T + U$ = a constant, action is only a function of position.

Walras borrowed these equations and made F the vector of the prices of a set of traded goods and q the vector of the quantities of those goods purchased. He then defined the integral $\int F\,dq = T$ as the total expenditure on these goods, integrated the expression as an exact differential, and defined the scalar function of the goods x and y as $U = U(x, y, z)$. Amazingly, he concluded that the resulting scalar function represents or describes the "utilities" of those goods.

Walras assumed that these utilities, like the concept of potential energy in the physics, are unobservable and that their existence can only be "inferred" through linkage with observable variables. He then argued that relative prices are equal to the ratios of the marginal utilities of goods by defining the "potential field" of utility as the locus of the set of constrained extrema. Although the other creators of neoclassical economic theory treated utility as a derived phenomenon by viewing the utility field as the exogenous data to which market transactions adjusted, they used the same mathematics. The assumption that this market system is reversible and without history did not seem totally unreasonable because the second law of thermodynamics, the entropy law, had not been formulated.

4. Robert Nadeau, *Wealth of Nature*, pp. 37–77.
5. William Stanley Jevons, *The Principles of Science*, 2nd ed. (London: Macmillan, 1905), pp. 735–736.
6. Jevons, *Principles of Science*, p. 736.
7. Bruna Ingrao and Giorgio Israel, *The Invisible Hand: Economic Equilibrium in the History of Science* (Cambridge, Mass: MIT Press, 1990), p. 97.
8. Léon Walras, "Letter to Louis Ruchonnet," in *Correspondence of Léon Walras and Related Papers*, ed. William Jaffé (Amsterdam: North-Holland, 1965) vol. 1, p. 201.
9. Léon Walras, *Elements of Pure Economics* (New York: Kelly Watson, 1960), p. 63.
10. Walras, *Elements of Pure Economics*, p. 69.
11. Walras, *Elements of Pure Economics*, p. 40.
12. Francis Ysidro Edgeworth, *Mathematical Physics* (London: Routledge, 1881), pp. 9, 12.
13. Vilfredo Pareto, *Manual of Political Economy* (New York: Augustus M. Kelley, 1971), pp. 36, 113.
14. Robert Nadeau, *The Environmental Endgame: Mainstream Economics, Ecological Disaster, and Human Survival* (New Brunswick, N.J.: Rutgers University Press, 2006), pp. 81–145.
15. Nick Hanley, Jason E. Shrogren, and Ben White, *Environmental Economics in Theory and Practice* (New York: Oxford University Press, 1997), p. 358.

16. W. Michael Hanneman, "Valuing the Environment through Contingent Value," *Journal of Economic Perspectives* 8 (Fall 1994): 19.
17. Jerald J. Fletcher, Wiktor L. Adamowicz, and Theodore Graham-Tomasi, "The Travel Cost Model of Recreation Demand: Theoretical and Empirical Issues," *Leisure Studies* 12 (1990): 119–147.
18. Mark Sagoff, "Some Problems with Environmental Economics," *Environmental Ethics* 10 (Spring 1988): 55.
19. Robert C. Mitchell and Richard T. Carson, "Valuing Drinking Water Risk Reduction Using the Contingent Valuation Methods: A Methodological Study of Risks from THM and Giardia," paper prepared for Resources for the Future, Washington, D.C., 1986.
20. George Tolley et al., "Establishing and Valuing the Effects of Improved Visibility in Eastern United States," paper prepared for the Environmental Protection Agency, Washington, D.C., March 1, 1984, http://yosemite.epa.gov/ee/epa/eerm.nsf/vwAN/EE-0003-01.pdf/$file/EE-0003-01.pdf.
21. James Bowker and John R. Stoll, "Use of Dichotomous Choice Nonmarket Methods to Value the Whopping Crane Resource," *American Journal of Agricultural Economy* 23 (May 1987): 372–381.
22. Kevin J. Boyle and Richard C. Bishop, "Valuing Wildlife in Benefit-Cost Analyses: A Case Study Involving Endangered Species," *Water Resources Research* 23 (May 1987): 943–950.
23. William D. Norhaus, "Reflections on the Economics of Climate Change," *Journal of Economic Perspectives* 7 (Fall 1993): 14.

CHAPTER 8

1. "Hard Times about the Bailout," *New York Times*, Sept. 10, 2008.
2. Ellen Brown, "How Brokers Became Bookies: The Insidious Transformation of Markets into Casinos," July 12, 2010, www.webofdebt.com/articles/brokers_bookies.php.
3. Milton Friedman, *Capitalism and Freedom* (Chicago: University of Chicago Press, 1982).
4. Milton Friedman, *Free to Choose* (New York: Harcourt Brace Jovanovich, 1980).
5. Thomas L. Friedman, "Foreign Affairs; Time of the Turtles," *New York Times*, Aug. 15, 1998.
6. Thomas L. Friedman, *The Lexus and the Olive Tree* (New York: Random House, 2000), pp. 87–88.
7. Friedman, *Lexus and the Olive Tree*, pp. 298, 302.
8. Francis Fukuyama, *The End of History and the Last Man* (New York: Penguin, 1992). p. 1.
9. Fukuyama, *End of History*, pp. 2–3.
10. George Gilder, *The Spirit of Enterprise* (New York: Simon & Schuster, 1984).
11. George Gilder, *Wealth and Poverty* (New York: Basic Books, 1981), p. 80.

12. Kevin Kelly, *Out of Control: The New Biology of Machines, Social Systems, and the Economic World* (Reading, Mass: Perseus, 1994).
13. Robert Samuelson, "Markets 'R' Us," *Newsweek*, Dec. 20, 1998.
14. Robert Bartley in *Backward and Upward: The New Conservative Writing*, ed. David Brooks (New York: Vintage, 1996), pp. 197–198.
15. Harvey Cox, "The Market as God: Living in the New Dispensation," *Atlantic*, March 1999.
16. David R. Loy, "The Religion of the Market," in *Worldviews, Religion, and the Environment*, ed. Richard C. Foltz (Toronto: Wadsworth, 2003), p. 67.
17. Loy, "Religion of the Market," p. 73.
18. Robert H. Nelson, *Economics as Religion: From Samuelson to Chicago and Beyond* (University Park: Pennsylvania State University Press, 2002).
19. Herman E. Daly and Kenneth N. Townsend, eds. *Valuing the Earth: Economics, Ecology, Ethics* (Cambridge, Mass.: MIT Press, 1993), p. 21.
20. Herman E. Daly, "Allocation, Distribution, and Scale: Toward an Economics That Is Efficient, Just, and Sustainable," *Ecological Economics* 6 (Dec. 1992): 186.
21. Daly, "Allocation, Distribution, and Scale," pp. 190–191.
22. Herman Daly, *Steady State Economics* (Washington, D.C.: Island, 1991), p. 17.
23. Robert Nadeau, *The Environmental Endgame: Mainstream Economics, Ecological Disaster, and Human Survival* (New Brunswick, N.J.: Rutgers University Press, 2006), pp. 172–179.
24. David C. Korten, "Taking Ecological Economics Seriously: It's the Biosphere, Stupid," keynote presentation to the U.S. Society for Ecological Economics, June 19, 2011, http://livingeconomiesforum.org/taking-ecological-economics-seriously.

CHAPTER 9
1. Ernst Mayr, *Toward a New Philosophy of Biology* (Cambridge, Mass: Harvard University Press, 1988), pp. 66–74.
2. Thomas Berry, *The Dream of the Earth* (San Francisco: Sierra Club Books, 1988), p. 87.
3. John Blofeld, trans., *The Zen Teaching of Huang Po: On the Transmission of Mind* (New York: Grove, 1970).
4. Eugene d'Aquili and Andrew B. Newberg, *The Mystical Mind: Probing the Biology of Religious Experience* (Minneapolis, Minn.: Fortress, 1999).
5. Henry James, ed., *The Letters of William James* (Boston: Atlantic Monthly Press, 1920), vol. 2, pp. 149–150.
6. Robert Kane, *Though the Moral Maze: Searching for Absolute Values in a Pluralistic World* (New York: Paragon House, 1998); Jeffrey Wattles, *The Golden Rule* (New York: Oxford University Press, 1996).
7. Roger S. Gottlieb, *A Greener Faith: Religious Environmentalism and Our Planet's Future* (New York: Oxford University Press, 2006), p. 71.
8. E. O. Wilson, *Biophilia* (Cambridge, Mass.: Harvard University Press, 1984), p. 145.

9. Eric Fromm, *The Art of Loving* (New York: Harper, 1989), p. 100.
10. Wendell Berry, *The Gift of Good Land* (San Francisco: North Point, 1981), p. 281.
11. John Burroughs, *The Gospel of Nature* (Bedford, Mass: Applewood, 1995), p. 32.
12. Michael S. Gazzaniga, *Human: The Science behind What Makes Us Unique* (New York: Harper Perennial, 2008), p. 148.
13. Rachel Carson, *The Sense of Wonder* (New York: Harper, 1984), p. 101.
14. Aldo Leopold, "Excerpts from *A Sand County Almanac*," in *Worldviews, Religion, and the Environment*, ed. Richard C. Foltz (Toronto: Wadsworth, 2003), p. 433.
15. William James, *The Varieties of Religious Experience* (New York: Longmans, Green & Co., 1902), p. 367.
16. David Orr, *Earth in Mind: On Education, Environment, and the Human Prospect* (Washington, D.C.: Island, 2004), p. 210.
17. Václav Havel, *Disturbing the Peace* (New York: Vintage, 1991), p. 181.
18. International Energy Agency (IEA), *World Energy Outlook 2011*, executive summary, www.iea.org/Textbase/npsum/weo2011sum.pdf.
19. Fatih Birol quoted in Fiona Harvey, "World Headed for Irreversible Climate Change in Five Years, IEA Warns," *The Guardian*, Nov. 9, 2011, www.guardian.co.uk/environment/2011/nov/09/fossil-fuel-infrastructure-climate-change.
20. IEA Launch of *World Energy Outlook 2011*, press conference held in London, Nov. 9, 2011, www.iea.org/newsroomandevents/news/2011/november/name,19882,en.html.
21. United Nations Framework Convention on Climate Change, Status of Ratification of the Convention, March 21, 1994, unfccc.int/essential_background/convention/status_of_ratification/items/2631.php.
22. For the National Religious Partnership for the Environment, see www.nrpe.org; for the Earth Charter Commission, see earthcharter.org.
23. Roger S. Gottlieb, *A Greener Faith: Religious Environmentalism and Our Planet's Future* (New York: Oxford University Press, 2006), p. 117.

INDEX

Africa
   climate change, 4–5
   complex language systems, 12, 26
   Save the Children Foundation, 32
African Savanna, competition, 71
African slaves, trade, 39
Against Mechanism, Morowski, 166*n*.3
agricultural revolution, Fertile
      Crescent, 142
agriculture
   dependence on oil, 41
   European farming, 38–39
   global food production system, 39
   global system, 41
   mechanization, 39
Amazon rainforest, 21
American economy, fossil fuels, 3
American International Group (AIG), 122
American news media
   dire environmental predictions, 3
   natural disasters, 2
American Recovery and Reinvestment
      Act, Obama, 121
Americas, European farming in, 38–39
Andes, melting glaciers, 21
Antarctica, melting ice sheets, 21
Apollo 11 spacecraft, 62, 137
Aquinas, Thomas, 78
Arctic, increasing average
      temperature, 19
Arctic glaciers, global warming, 4
Argentina, 39
Aristotle, 78
Arrow, Kenneth, 93, 119
Asia
   climate change, 1, 2, 5

history of human family, 34–35, 38
human survival, 20
Aspect, Alain, 58
Aspect experiments, 58
Augustine, 78, 129
Australia, 4, 34, 35, 36, 39, 41

Bangladesh, 5, 85
banking systems and climate change, 21
Bartley, Robert, 128
BBC surveys, 23
behavioral sciences, brains, 43
Bell, John, 57, 141, 161*n*.14
Berry, Thomas, 10, 56, 141
Berry, Wendell, 146
Bhagavad Gita, 143
Bible, language translations, 82
biology, 11
   alliance between organisms, 67–68
   competition vs. cooperation, 69–71
   life on other planets, 138
   modern synthesis, 64, 65
   new story, 60, 64, 73–75
   parts and wholes in public policy,
      72–73
   parts and wholes in system of life,
      66–69
biosphere, human transformation of,
      36–38
Birol, Fatih, 151
bloody wars, 81–82
Book of Revelation, 143
brain
   complex language, 29
   imaging systems, 28, 144
   neuroscience, 28–29, 43

Broca's area, 28, 31
Brown, Donald, 33
Brown, Peter, 9
Buddhism, 143, 144
Buddhists, 153–154
Burroughs, John, 146
Bush, George W., 22, 87, 88, 121, 164n.22

call and put options, weather-based, 124
Calvin, John, 81
capitalism, 128
Capitalism and Freedom, Friedman, 126
carbon dioxide concentrations
  impact on global warming, 17
  predictions, 20
carnivores, competition, 70–71
Carson, Rachel, 147
Catholic Church, 77–79, 80
Catholic nuns, 144
cell, alliance between mitochondria and, 67
Ch'an Master Huang Po, 143
chaos theory, 18
children, language, 29–30
China
  cross-border conflicts, 6
  drinking water shortage, 2
  flooding and mudslides, 2
  monsoons, 5
Chirac, Jacques, 10
Christianity, 145
churches of state, emergence of, 80–83
cities, death rates exceeding births, 40
classical economics, 93, 105, 107
classical physics, religion, 45–46
climate change
  Africa, 4–5
  campaigning against truths of science, 147–148, 154–155
  contrarians, 23, 24
  environmental economists, 118–119
  environmental scientists, 138
  false gods and politics of, 149–155
  nonlinear systems theory, 18
climategate, alleged conspiracy, 23
climate system
  computer predictions, 72
  global market, 3

global warming triggering irreversible changes, 21
irreversible large-scale change, 21–22
clockwork universe, 93
Cold War, 127, 139
Colinvaux, Paul, 71
collateralized debt obligations (CDOs), 122
Columbian Exchange, 39
Commodity Futures Modernization Act, 122
communication, worldwide network, 42
communism, 128
compassion, limbic system, 32, 33
competition for resources, African Savanna, 71
competition for survival, Darwin, 69–71
complex dynamics, 18
computational tools, nonlinear systems theory, 65
computer systems, climate models, 72
contingent valuation surveys, environmental economists, 117–118
Convention on Biological Diversity, United Nations, 86
Convention on the Law of the Sea, United Nations, 86
Convention on the Non-Navigable Uses of International Watercourses, United Nations, 86
Convention to Combat Desertification, United Nations, 86
cooperation, competition vs., 69–71
coral reefs, global warming, 20
Corn Belt, United States, 39
corn laws, Ricardo, 103
cosmos
  classical physics, 46
  conscious universe, 74–75
  physical reality, 59–61
  as 3-D movie, 51–52
Cosmos, Sagan, 138
Cox, Harvey, 121, 129–130
crowd diseases, lethality, 39–40
Cummings, E. E., 43

Daly, Herman, 131–133
Darwin, Charles, 25
  competition for survival, 69–70

elephant calculations, 69–70
evolution, 63–66
natural selection, 67
degradable materials, environment, 40
deism
   classical physics and existence of God, 92–93, 104
   invisible hand, 105
delayed-choice thought experiment, Wheeler, 54–55
democratic capitalism, 128
Democratic Party, 88
deregulation, derivatives market, 122
derivatives trading
   hedging against unpredictable weather, 123–124
   religion of market, 135–136
derivative trading, 122
Der Spiegel, 23
d'Espagnat, Bernard, 58
differential calculus, economics and physics, 111
Dillard, Annie, 33, 45
dinosaurs, end of age, 24, 35
disinformation campaigns
   fossil fuels, 3
   global warming, 4
DNA, life, 66–69
Dobrianski, Paula, 88
Dobzhansky, Theodosius, 64
Dodd Frank bill, 123
donations, Save the Children Foundation, 32
dream, humane and equitable world, 139–140, 147
The Dream of Earth, Berry, 10
Durban Platform for Enhanced Action, 152
Dyson, Freeman, 59

Earth Charter Commission, 153
Earth Summit, United Nations, 86
East Africa, climate change, 5
Ecole des Mines, 111
ecological economics, 130–136
ecology
   Fertile Crescent, 35–36
   human life, 146
economic narrative

   geopolitical reality, 73–74
   market systems, 13
economic recession, financial meltdown (2008), 3, 120, 121–123
economics
   assumptions, 114–115
   basic goals, 132
   competitive prices, 113
   creators of neoclassical, 107, 110–115
   differential calculus, 111
   environmentally responsible theory, 135
   global system, 133
   goods consumption in market systems, 111–112
   invisible hand, 99–100
   Keynesian, 125
   natural laws of, 93, 97–99, 124, 153
   new theory of, 130–136
   physics, 107–108
   tightening invisible chains, 100–104
Economics as Religion, Nelson, 130
economic stimulus plan, Bush, 121
economic system, connections in global, 25–26, 41–42
ecosystem
   evolved mechanisms, 71
   global economy, 132–135
Edgeworth, Francis Ysidro, 107, 114
Einstein, Albert
   general theory of relativity, 48–49
   human body, 59
   physical reality, 57
   quantum theory, 58
   special theory of relativity, 47–49
Elements of Pure Economics, Walras, 112
elephants, Darwin, 69–70
El Nino, 21
emergence, 64
   ancestor of DNA, 66
   churches of state, 80–82
   environmental ethos, 146–149
   intelligent life-forms, 138
   manufacturing techniques, 104
   matter quanta, 56
   new story of science, 66–69
   religious environmentalism, 8, 135–136, 140–141
   telecommunications, 42
emotions, mirror neuron system, 31

empathy, limbic system, 32, 33
The End of History, Fukuyama, 128
energy
    mathematical formalism, 109
    neoclassical economic theory, 111, 131–132
    physics, 107–108
    population, 38
    sun, 68
energy tax, 3
entropy, physics, 131
environmental crisis
    conscious universe, 74–75
    existence of false gods, 152
    neuroscience, 43
    political debate about, 89–91
    pollution, 118
environmental economists
    climate change, 118–119
    contingent valuation surveys, 117–118
    cost-benefit analyses, 117
    invisible hand, 116
    neoclassical economic theory, 116–117
environmentalism
    Carson, 147
    new ethos, 146–149
    prophets warning of natural disaster, 148–149
    religious, 8, 14, 140–141, 146, 153–154
environmentally concerned people, religious traditions of world, 8, 13–15
environmental predictions, news media, 3
environmental problems
    invisible hand, 115–120
    scientifically viable economics, 152–153
    United Nations agreements, 86
    worldwide movement, 140–141
Environmental Protection Agency (EPA), 88
environmental scientists, climate change, 138
environmental subsystems, irreversible changes, 6, 21
environmental systems
    human activity and, 10
    relationship between human and, 7
Epstein, Seymour, 31
An Essay on the Principle of Population as It Affects the Future Improvement of Society, Malthus, 101
ethical behavior, moral circle, 89–91
Ethiopia, Save the Children Foundation, 32
ethnicity, 11
European farming, agriculture in Americas, 38–39
evolution
    biological life on other planets, 138
    Darwin, 63–66
    language, 29–30
    spontaneous moral behavior, 33
evolved behavior, competition, 71

Faraday, Michael, 110
fascism, 128
feedback loops
    biological systems, 142
    DNA codes for enzyme production, 68
    grammar, 30
    limbic system, 33
    self-reinforcing positive, 18, 20
    suffering of individuals, 33
Fertile Crescent
    agricultural revolution, 142
    agricultural system, 38
    domesticated plants and animals, 37–38
    ecology, 35–36
    energy in, 38
    human transformation of biosphere, 37–38
Feynman, Richard, 45, 52
financial markets, meltdown (2008), 3, 120, 121–123
Fitch, 122
flooding, United States, 21
Florentine Council, 79
food staples, global human population, 39
force field of gravity, space-time, 48, 49
Forum on Religion and Ecology, Yale University, 15, 153
fossil fuels

disinformation campaigns, 4
transportation, 39
fossil record, energy of sun, 68
fossil remains, artifacts in, 27
Framework Convention on Climate Change, United Nations, 86, 170n.21
Free to Choose, Friedman, 126
French Calvinists, 81
Friedman, Milton, 125–126
Friedman, Thomas, 127
Fromm, Erich, 146
Fukuyama, Francis, 128
functional magnetic resonance imaging (fMRI), brain imaging, 28

Galbraith, John Kenneth, 107
Gandhi, Mahatma, 147
Gazzaniga, Michael, 147
General Assembly, United Nations, 84
Genesis, 129, 130
genetic inheritance, Mendel, 64
geopolitical climate, securing resources, 6
geopolitical reality, 73–74, 76
German people, nationalism, 83–84
Germany, conflicts between nation-states, 5
Gilder, George, 128
Gisin, Nicolus, 58, 161n.14
Gisin experiments, particles, 58, 161n.14
global agricultural system, complexity, 41
global demand for oil, population, 40
global derivatives market, value, 123
global economy
connections, 25–26, 41–42
meltdown of financial markets (2008), 3, 120, 121–123
global food production, agriculture, 39
global market system
climate, 3
economy, 73
fossil fuel-based, 120, 134
global population, demand for oil, 40
global thermohaline conveyor, 4
global thermohaline system, collapse, 6
global warming. *See also* climate change
bad news about, 19–20
carbon dioxide concentrations, 16–17, 20
conditions resulting from, 1–2
derivatives trading, 123
disinformation campaigns, 4
environmental impact on market systems, 123–124
false gods and politics of, 149–155
international government, 89–90
man-made, and natural disasters, 2
news media, 15–16
politics of, 22–24
public debates, 72–73
satellites and ground observation systems, 17, 19–20
science of, 15–18
World Meteorological Organization (WMO), 1–2
Gödel, Kurt, 59
Godwin, William, 101
Goldwater, Barry, 126
Gottlieb, Roger, 14, 145, 154
Gould, James, 70
Greece, blocking marine pollution agreement, 85
greenhouse gases
carbon dioxide, 20
emissions, 14, 16, 24, 41, 150–151
Group of Eight, 87
international treaties, 87–88, 151–152
manufacturing processes, 131
methane, 19
Greenland Ice Sheet, 2, 21
Grim, John, 15
gross national product (GNP)
health of economy, 73
market economies, 115
ground observation systems, global warming impacts, 18, 19–20
Ground of Being, 143, 144
Group of Eight, greenhouse gas emissions, 87
Gulf Stream, circulation pattern, 4

Hahn, Hans, 93, 119
Hamilton, William, 166n.3
Hansen, James, 20
Harvard Divinity School, 14
Harvard University, 126

Harvard University Press, 15
Havel, Václav, 149
heat waves, record-breaking, 1
Heisenberg, Werner, 46, 74
Helmholtz, Hermann-Ludwig Ferdinand von, 107
herbivores, competition, 71
hereditary monarchy, 128
Himalayan-Tibetan plateau, melting glaciers, 21
Himalayas, disruptions in water flow, 2
Hinduism, 143, 145
History of the German People, Luden, 83
Hitler, Adolf, 83
Holy Roman Empire, 81, 83
Hominids, brains and bodies of, 26
Hoover Institution at Stanford, 126
Huguenots, 81
human beings, slave trade, 39
human body, Einstein, 49
human brain, 28
  complex language systems, 29, 138
  social science model, 42
human family
  brief early history of, 34–36
  mutations in genes, 34
human future, market consensus, 127–129
human genetics, 11
human history
  invisible hand, 99
  religion of market, 24, 129–130
human identity, 7, 13
humanity
  artifacts with fossil remains, 27
  brains and bodies of hominid ancestors, 26
  complex human societies, 27
  dream of enlightened future, 139–140
  early history of human family, 34–36
  environmental crisis, 42–43
  human transformation of biosphere, 36–38
  language, 29–30
  neuroscience of brain, 28–29
  new view of human nature, 31–34
human nature, new view of, 31–34
human population

dream of humane and equitable world, 139–140
  energy regime of oil, 38–42
human settlers, Fertile Crescent, 35–36
human survival, new terms of, 20–22, 140
human transformation, biosphere, 36–38
human will, force of, 113
hunter-gatherer tribes
  fairness and reciprocity, 34
  moral behavior, 32

Ice Age, 39
idealists, dream of humane and equitable world, 139–140
illness, lethality of crowd diseases, 40
incompleteness theorem, Gödel, 59
India
  cross-border conflicts, 6
  drinking water shortage, 2
  migration and wars, 21
Industrial Revolution
  carbon dioxide concentrations, 17
  European population, 39
infants, language, 29–30
Institutes of the Christian Religion, Calvin, 81
Interfaith Climate Change Network, 153
Intergovernmental Panel on Climate Change (IPCC), 17, 22, 86
International Energy Agency (IEA), 150–151, 170n.18
international government
  global warming, 89–90
  present system of, 84–86
international treaties, 6, 87, 88, 150–151
invisible hand
  businessman, 99–100
  deism metaphor, 105
  environmental problems, 115–120
  market consensus, 125–127
  natural laws, 93, 94–95, 101–103
  necessities of life, 97
Islam, 144, 145

Jackson, Tim, 134
James, William, 145, 147

Japan
  blocking marine pollution agreement, 85
  climate change campaign, 153
  conflicts between nation-states, 5
  Kyoto agreement, 151
  population and energy, 38, 40, 41, 85
Jevons, William Stanley, 107, 110–112
Jewish synagogues, 153
Joule, James Prescott, 110
Judaism, 145
Judeo-Christian God, deism, 92–93, 104, 105
Judeo-Christian tradition
  false god of market, 135
  religion of market, 129
  science and religion, 143–144

Kantian philosophy, 111
Kelly, Kevin, 128
Keynes, John Maynard, 125
King, Martin Luther Jr., 147
Korten, David, 134
Kyoto Protocol, 86
Kyoto Treaty, 88
landlords, fertile land as gift of nature, 103–104

language
  human brain, 28
  human species, 29–30
language systems
  barriers, 43
  Ethiopia, Kenya and Tanzania, 26
  evolution, 138, 142
  human brain, 29
  modern humans, 12
La Nina, 2
Laplace, Pierre-Simon de, 50
law of gravity
  forces regulating prices, 113
  Newton, 93
Laws, Richard, 70
learning
  human brain, 28
  mirror neuron system, 31
Lehman Brothers, 122
Leopold, Aldo, 147

The Lexus and the Olive Tree, Friedman, 127
life. *See also* biology
  part and wholes in system of, 66–69
  limbic system, empathy, sympathy and compassion, 32, 33
Linnean Society, 63
Lord Kelvin, 110
Lord Krisna, 143
Loy, David, 130
Luden, Heinrich, 83
Luther, Martin, 80
Lutherans, 153
Lutz, Frank, 22

McGill University, 9
Machiavelli, Niccolò, 79
machines, mind and nature, 97–99
McKibben, Bill, 14
maize, food staple, 39
Malthus, Thomas, 92, 100–104
manifest destiny of U.S., new state religion, 127–129
man-made global warming, 2
Margulis, Lynn, 66–67
market consensus
  human future in, 127–128
  manifest destiny of U.S., 127–129
market systems
  competitive prices, 113
  consumption of goods, 111–112
  economic narrative, 13
  environmental impact of global warming, 123–124
  invisible hand and market consensus, 125–127
  religion of market, 24, 129–130
  Ricardo, 103
Maronite Christians, 153
mass migration, starving, desperate people, 21
mathematics
  economics, 114–115
  neoclassical economic theory, 108–109, 134
matter, classical physics, 46
Mayr, Ernst, 62, 64
Medawar, J. S., 64
Medawar, P. B., 64
memory, human brain, 28

Mendel, Gregor, 64
metaphysics
  cosmic machine and hidden chains, 97–99
metaphysics (*Cont.*)
  invisible hand, 96
  Smith, 104–105
  theory of reality, 13
Mexico, tension with U.S., 6
migration, original family of modern humans, 34
mirror neurons, emotions and behavior, 31
mitochondria, alliance with cell, 67
modern synthesis, biology, 64, 65
Moody's, 122
moral behavior
  brains, 42–43
  environmentalism, 147
  spontaneous, 32, 33
moral circle, principles and behavior, 89–90
moral philosophers, deism, 92–93
Mother Theresa of Calcutta, 33

NASA, 20, 149–150
national identity, 73, 77, 82–84, 147
nationalism, 11, 73, 77, 82–84, 147
National Religious Partnership for the Environment, 153, 170*n*.22
Nation-states
  conflicts between, 5
  political narrative, 13
  political power, 10–11
  sovereign, 73
  United Nations, 84
natural, term, 113
natural disasters
  news media, 3
  scientific prophets warning, 148–149
natural laws
  economics, 93, 98, 114–115, 124, 153
  human beings, 99–100
  invisible hand, 93, 94–95, 101–103
  physical and, 96
  supply and demand, 104
  unequal powers, 102
natural liberty, system of, 95–97, 98
natural resources, neoclassical economics, 132–133

natural selection, Darwin, 67
Nature, 20
nature, fertile land as gift of, 103–104
Nazi church of state, 84
Nelson, Robert, 130
neoclassical economics, 125, 166*n*.3
  assumptions, 114–115
  consumption of goods, 111–112
  creators of, 107, 110–115
  ecological economics, 131
  energy, 131–132
  mathematical formalism, 109–110, 134
  natural resources, 132–133
  physics theory, 108
  replacing, 153
neocortex, language functions, 28
Netherlands, storms, 5
neuroscience, brain, 28–29, 43
news media, global warming, 15–16
Newsweek, 129
Newtonian physics, 93, 111
  religion, 45–46
New York Times, 88, 121, 127
Nietzsche, Friedrich, 74
Ninety Five Theses, Luther, 80
nongovernmental organizations (NGOs), 17
nonlinear systems theory
  climate change, 17–18
  computational tools, 65
nonverbal system, spontaneous moral behavior, 32
Norway, 85

Obama, Barack, 23, 88, 121
ocean dumping of radioactive waste, 86, 164*n*.15
oil and energy
  disinformation campaigns, 4
  global population, 40
  transportation, 39, 41–42
On the Principles of Political Economy and Taxation, Ricardo, 103
The Origins of Species, Darwin, 63
Orr, David, 41, 149
Orwellian lies, 148, 154

Out of Control: The New Biology of Machines, Social Systems and the Economic World, Kelly, 128

Pakistan, 1, 2, 6, 21
Pareto, Vilfredo, 107, 114
particle interactions, fact of nature, 56–59, 141–142
particles, quantum entanglements, 59
Pauli, Wolfgang, 60
Peace of Westphalia, 81–82
Pentagon, 4
    environmental crisis, 7
    Policy Analysis Market (PAM), 123
petrochemical industry, 40
Pew Research Center, 23
phenomena of mind, 110
philosophy
    climate change campaigns, 147–148
    system of natural liberty, Smith, 95–97
physical theories, energy, 107–110
physics, 11, 43
    classical, 46
    differential calculus, 111
    economics theory, 107–108
    energy, 113
    entropy, 131
    general theory of relativity, 48–49
    lessons from new story, 59–61
    natural laws, 96
    neoclassical economics, 166n.3
    new fact of nature, 56–59
    new story of, 46
    Newtonian, 93, 111
    quantum, 49–56
Planck, Max, 48
plant life, Fertile Crescent, 35–36
Policy Analysis market (PAM), 123
political ideology, 11, 43, 128
political narrative, sovereign national-state, 13
political power
    nation-states, 11, 89, 152
    United Nations, 7, 73, 77, 84
politics
    decisions in economics planning, 119
    global warming, 22–24, 149–155
pollution
    campaign, 153
    economics, 118
    externalities, 116–117
    manufacturing, 41, 131
    marine, 84–86
population growth
    elephants, Darwin, 69
    food supply, 101
    human growth and energy, 38–42
    unseen chains of the laws of, Malthus, 102–103
positron emission tomography (PET), brain imaging, 28
post-Kyoto agreement, 87, 88, 151
potatoes, food staple, 39
potential energy
    Hamiltonian principle, 167n.3
    prices for goods, 166–167n.3
prices, market system, 113
Prigogine, Ilya, 65
The Prince, Machiavelli, 79
The Principles of Science, Jevons, 110
Prodi, Romano, 88, 165n.23
prophets, warning of natural disaster, 148–149
Protestant Reformation, 77, 80–82
public policy
    invisible hand, 99–100
    life as system, 72–73
public trust, politics of global warming, 23–24
Pygmies, 35

quantum physics
    collapse of wave function, 51, 52
    delayed-choice thought experiment, 54–55
    fact about particles in nature, 56–59, 140–141
    Pauli's contribution, 60
    quantum entanglements, 59
    relativistic quantum field theory, 49–50
    Schrodinger wave equations, 50, 51, 52
    self and world, 49–56
    two-slit experiment, 52–54
    universe as 3-D movie, 51–52
    wave-interference patterns, 50, 51
    wave-particle dualism, 50, 51, 142
    Wheeler's prediction, 55
Qur'an, 144

radioactive waste, ocean dumping, 86, 164*n*.15
rating agencies, 122
Reagan, Ronald, 126
reality, industry and market systems, 113
relativity theory, 161*n*.14
　general, 48–49
　special, 47–49
religion. *See also* state religion
　beliefs, 43
　beliefs and practices, 11
　Catholic Church, 77–79
　dialogue between truths of science and, 8, 11, 12
　economists and absentee deistic god, 119
　environmentally concerned people, 8, 13–15
　false gods of global warming, 149–155
　invisible hand, 93, 94–95
　Newtonian or classical physics, 45–46
　strict division of science and, 60
　truths of science and, 143–146
religion of the market, 24, 129–130, 135–136, 154
　in market we trust, 127–129
religious environmentalism, worldwide movement, 8, 14, 140–141, 146, 153–154
Republican primary elections, 135
residual warming, resulting conditions, 2–3
resources, geopolitical climate and securing, 6
Revelation, science and religion, 143–144
Ricardo, David, 92, 100, 103–104
Rizzolatti, Giacomo, 31
Rocky mountains, melting glaciers, 21
Royal Institution, 110
Rucker, Rudy, 60
Russia, 2, 5, 39

Sagan, Carl, 138
Sagan, Dorion, 66–67
satellites
　global warming impacts, 17, 19
　photographs and videos, 11
Save the Children Foundation, 32
Schröder, Gerhard, 88
Schrodinger, Erwin, 59, 60
Schrodinger wave equations, quantum physics, 50, 51, 52
Schulze, Hagen, 83
science
　climate change campaigning, 147–148, 154
　global warming, 15–18
　warning of natural disaster, 148
science and religion
　new dialogue, 8, 11, 12
　strict division, 60
　truths of, 143–146
Search for Extraterrestrial Intelligence (SETI) project, 138
Security Council, United Nations, 84
self-organization, spontaneous, 65
Shakespeare, William, 25, 30
Shintoists, 153
Silk Road, trade, 38
Singer, Peter, 76, 89, 90
single photon emission tomography (SPEC), 144
Smith, Adam, 92
　deist, 104–105
　invisible hand, 116
　natural laws and markets, 102
　natural laws of economics, 93, 124
　"Of the Origin of Philosophy," 95
　The Theory of Moral Sentiments, 94, 96–97, 99
　The Wealth of Nations, 93, 94, 95, 96, 99
society
　natural laws and invisible hand, 99–100
　system of natural liberty, 95–97
Society for Ecological Economics, 134
South Africa, 21, 39
　food production, 39
　human survival, 21
South Central Pakistan, 1
South Korea, 5
sovereign nation-states, 73, 76–77, 153
　emergence of churches of state, 80–82

international government, 84–86
legal principle of state sovereignty, 87–88
origins of construct of, 77–79
post-Kyoto agreement, 151
Soviet Union, 84, 126
space-time coordinates
  gravitating mass, 48, 49
  special theory of relativity, 47–48
speed of light, special theory of relativity, 47
Speth, Gus, 134
The Spirit of Enterprise, Gilder, 128
spiritual reality, human life and consciousness, 11–12
spontaneous self-organization, 65
Standard & Poor's (S&P), 122
standard social science model, 42
Stapp, Henry, 52
state religion. *See also* religion
  manifest destiny of U.S., 127–129
  religion of market, 24, 129–130, 135–136, 154
state sovereignty, legal principle of, 87–88
Steady State Economics, Daly, 132
Stinchcombe, Arthur, 79
stock markets, climate system, 21–22
Summary for Policy Makers, Post Kyoto agreement, 87
summer of 2010, heat waves, 1
sun, energy, 68
super-family, 78
supply and demand, laws of, 104
Swimme, Brian, 56
symbiotic alliances, organisms, 67
sympathy, limbic system, 32, 33
systems, machines, Smith, 97–98
systems dynamics, 18

Taoism, 143, 145
Tao te Ching, 143
Tea Party, 129
telecommunications, worldwide network, 42
Tellus Institute, 153
Theory of Everything, 60
theory of fractals, 18
The Theory of Moral Sentiments, Smith, 94, 96–97, 98

Theory of Political Economy, Jevons, 110
thermodynamics
  entropy, 131
  laws of, 131, 132
Third Reich, 84
Thomson, William, 110
Tibetan Plateau, 2
Time Magazine, 129
transportation, fossil fuel-based, 41, 42
Tucker, Mary Evelyn, 15
two-slit experiment, quantum physics, 52–54

United Church of Christ, 153
United Nations, 77, 82
  international government, 84–86
  political power, 7
United Nations Population Fund, 41
United States
  climate disruptions, 5
  economy and climate, 118–119
  flooding, 21
  food and water shortages, 21
  Kyoto Treaty, 88
  manifest destiny of, 127
  marine pollution agreement, 85
  securing resources, 6
universe
  as 3-D movie, 51–52
  conscious, 74–75
  new story of, 141–143
University of Chicago, 126
University of East Anglia, 22
University of Geneva, 58
University of Glasgow, 94
University of Paris-South, 58
University of Pennsylvania, 144
U.S. Congress, global warming, 73

Victor, Peter, 134

Wall Street Journal, 129
Walras, Léon, 107, 111–112, 166–167n.3
warped space-time, gravitating mass, 48, 49
wars, starving, desperate people, 21
wave-interference patterns, quantum physics, 50, 51
wave-particle dualism, quantum physics, 50, 142

Wealth and Poverty, Gilder, 128
The Wealth of Nations, Smith, 93, 94, 95, 96, 99, 104–105
Weber, irreplaceable cultural values, 78
Wernicke's area, 28
Western metaphysical tradition, 13
Wheeler, John A., 54
Whitman, Christine Todd, 88, 164*n*.21
Whitman, Walt, 10
Wilbur, Richard, 137, 139
Wilson, E. O., 146
World Bank, 3
World Council of Churches, 153

World Energy Outlook 2011, 150–151, 170*n*.18
World Meteorological Organization, 1–2
World War II, 83, 84, 125
worldwide movement, religious environmentalism, 8, 14, 140–141, 146, 153–154
worldwide network, telecommunications, 42

Yale University, Forum on Religion and Ecology, 15, 153